A New Century of Biology

A New Century of Biology

Edited by W. John Kress and Gary W. Barrett

Foreword by Ernst Mayr

SMITHSONIAN INSTITUTION PRESS • Washington and London

In association with the
AMERICAN INSTITUTE OF BIOLOGICAL SCIENCES

Copy editor: Susan M. S. Brown
Production editor: Robert A. Poarch

Library of Congress Cataloging-in-Publication Data
A new century of biology / edited by W. John Kress and Gary W. Barrett.
     p. cm.
    Includes bibliographical references.
    ISBN 1-56098-984-X (alk. paper) — ISBN 1-56098-945-9 (pbk. : alk. paper)
    1. Biology. I. Kress, W. John. II. Barrett, Gary W.
    QH307.2.N49 2001
    570—dc21   2001018376

British Library Cataloguing-in-Publication Data is available

Manufactured in the United States of America
08 07 06 05 04 03 02 01   5 4 3 2 1

Contents

Foreword
Biology in the Twenty-First Century

ERNST MAYR

Most living biologists, particularly the younger ones, may not realize how young a science biology is. To discover this, let us take a glance at the history of biology. The field had a most promising beginning with Aristotle in the third century before Christ. Aristotle was quite extraordinary because he was not only an excellent naturalist, particularly a very knowledgeable student of marine animals, but also interested in physiology and embryology. Alas, this promising beginning was not further followed up for almost two thousand years. In the sixteenth, seventeenth, and eighteenth centuries the famous Scientific Revolution took place, characterized by names such as Galileo, Kepler, Newton, and Descartes. When they made their great discoveries in the physical sciences, where was biology?

To be sure, there was a considerable interest in living nature but no cohesive science. There were two camps: one of the naturalists, who studied nature in the spirit of natural theology, who discovered ever more pieces of evidence for God's near-perfect design of the living world. The writings of these naturalists reveal a remarkable understanding of the life histories and adaptations of living organisms but were not seen, at that time, as science. Medicine was the other stronghold of biology. It produced marvelous achievements in fields such as anatomy (Vesalius), embryology (Harvey), and physiology. Botany was vigorously advanced at that period because all medications were derived from herbs and exact identification of plants was of crucial practical value. All great botanists from the sixteenth to the end of the eighteenth century, with the single exception of John Ray, were medical doctors. This includes the great Linnaeus, often called the founder of systematics.

Around the year 1800 three different authors, the French zoologist Jean-Baptiste de Lamarck and two Germans, introduced the word *biology,* referring to the study of the world of life. These authors called for the development of such a science, but it did not yet exist. There could not be a science of biology until scientists had first learned a great deal more about the living world; the major biological disciplines had first to be founded. This took place in the thirty-eight years from 1828 to 1866: embryology (Karl Ernst von Baer, 1828), cytology (Theodor Schwann and Matthias Schleiden, 1830s), physiology (Claude Bernard and Hermann Helmholtz, 1840s), evolution (Alfred Russel Wallace and Charles Darwin, 1858–59), and genetics (Gregor Mendel, 1866). Yet a real synthesis did not occur until seventy-five years later. Until then the workers in functional biology (physiology, embryology) ignored the achievements of evolutionary biology (and genetics) and vice versa. Indeed, major controversies within each of these fields had to be settled first. As far as evolutionary biology is concerned, in the early 1930s there were two camps: the experimental geneticists, mostly interested in the mechanism of evolution and studying variation within a population as well as the achievement and maintenance of adaptation, and another camp consisting of the naturalists, systematists, and paleontologists, primarily interested in the study of biodiversity, that is, species, speciation, and macroevolution. Between 1937 and 1947 a synthesis of the two fields was achieved through a mutual understanding of each other's views. The result was the so-called evolutionary synthesis, actually very much a return to classical Darwinism, evolution as variation and selection.

The next major event was the founding of molecular biology through the discoveries of Oswald Avery and of James Watson and Francis Crick (1944–53). One might have expected that these discoveries would result in a major revolution in evolutionary biology, but this did not take place. Molecular biology made possible a fine-grained analysis, but it did not result in a refutation of the underlying Darwinian theory. Indeed, molecular biology made some magnificent contributions to our understanding of evolution, such as that the material of inheritance is nucleic acid rather than proteins, and that the genetic code is the same for all organisms from the bacteria up, indicating a single origin of life, but it did not touch the basic Darwinian framework. Perhaps the greatest contribution made by molecular biology was that it gave a new lease on life to developmental biology, which for several decades had been virtually dormant. One could refer to the last fifty years as the era of nucleic acids. But DNA gives only information and instructions, the real work in development is done by proteins. I foresee that the next fifty years will be more and more an era of the proteins.

Now, I will try to take a look at the future. With all these achievements, am I willing to say that the task of biology is finished? Not by any means! To be sure, the major theoretical framework of modern biology is remarkably robust. Even in a field as tumultuous as evolutionary biology, our currently accepted theories are remarkably similar to Darwin's original proposals. However, the basic philosophy of biology, as it has developed in the last fifty years, has become quite different from the classical philosophy of science as it prevailed from the Vienna school of Rudolf Carnap and Otto Neurath to Karl Popper and Thomas Kuhn. In the rejection on the one hand of all vitalistic theories and of concepts such as cosmic teleology, and on the other hand of all physicalist concepts such as essentialism, determinism, and reductionism, and their replacement by an acceptance of the frequency of random events, plural solutions, the importance of historical narratives, multiple causations, population thinking, and the greater significance of concepts than of laws in theory formations, the new biology has undergone a complete revolution.

But what about our understanding of the basic biological phenomena? I have the feeling that we are remarkably far advanced in this understanding, of let us say how a neuron works or the nature of genes. Where our knowledge lags behind is in the understanding of complex systems. The first of these is the developmental system, the development of a zygote from the fertilized egg to the finished adult. There is still an enormous amount to be learned, not only about the interaction of various kinds of genes, particularly regulatory genes, but also about the inducing interaction between different tissues.

The second complex system of which we still have a very insufficient understanding is, as we all know, the central nervous system. I wonder whether we will ever know all the interactions of the 3 billion neurons of our central nervous system, each of which has up to a thousand connections (synapses) with other neurons. I find it totally miraculous that at the age of ninety-five I can suddenly remember a name of which I had not thought for eighty years. There is no doubt that we still have a long way to go in our understanding of the mind.

The third complex system is the ecosystem, the interaction of the thousands of organisms from the largest trees to the smallest bacteria in the biota of a locality and what controls their presence, frequency, and interactions. There is a great future ahead in the study of these three systems.

What is the probability that the current advances in gene technology and other modern biological techniques will affect our personal lives? What is the chance that as a result of ongoing biological researches we may one day have to face something entirely new and undesirable? Unfortunately too many people take science fiction seriously and believe the most absurd accounts of possible future developments. They are afraid that the new gene techniques might produce monsters and other misfits. I am convinced that there is no chance for this whatsoever.

To be sure, there are now techniques for the replacement of defective genes, perhaps even in the germ line. I specifically said defective genes. At the present time we have neither the knowledge nor the methods to make people more intelligent or more altruistic or, for that matter, more stupid or more malicious. There is no subject on which in recent years more nonsense has been written than cloning. To talk about whole populations of Einstein clones is simply absurd. A single clone in a family might be justified occasionally, but this clone would not differ significantly from a monozygotic twin.

When thinking about the future of biology, let us think of the vast benefits biology has brought to humankind in the past. Indeed, biology is likely to continue to bring us equally unexpected benefits, particularly in medicine and agriculture. The great reduction of premature mortality in human beings, and correspondingly the virtual doubling of human life expectancy in the last one hundred years, is an achievement of biology.

Let me sum it up. Biology is healthy and well and is looking with confidence to great future achievements. Furthermore, and this is perhaps my most important conclusion, being a biologist is so much fun!

Preface

A New Century of Biology

W. JOHN KRESS AND GARY W. BARRETT

We have entered a "new century of biology" in which we expect to witness an explosion of discoveries that will revolutionize the biological sciences and in particular the relationship of human society to the environment. Just as Galileo, Copernicus, Newton, Einstein, and others made revolutionary breakthroughs in our understanding of the physical world and the universe, major conceptual breakthroughs in our understanding of the biological world are imminent. We have made outstanding progress in areas such as understanding the processes of evolution and ecology, discovering the biochemical structure of DNA, and making the simple but critical calculation of the magnitude of biodiversity on the planet. In the new century advances in the biological sciences will be made in a context of new technologies, in an environment changing rapidly because of human activities, and with a new relationship to global economics and social structure.

The voyages of discovery of the early naturalists during the eighteenth and nineteenth centuries were in many ways the beginning of the study of modern organismal biology. The exploration of unknown lands and habitats funded by monarchs and wealthy supporters led to the discovery of many new kinds of plants, animals, and microorganisms, both fossil and contemporary. The early naturalists focused on species and populations as well as the habitats in which these organisms were found. These broadly trained biologists studying the past as well as the present biota provided the raw data that were critical for the shaping of Darwin's theory of evolution by natural selection. A rigorous scientific explanation for the origin of complex biological systems had been formulated. Such organismal-based studies have continued to the present, resulting in the development of the fields of systematics, population biology, community ecology, and ecosystem or landscape-level sciences.

In the mid-1900s the biological community also experienced the emergence of a reductionist approach to studying life. The new biologists were strongly influenced by physicists and mathematicians, and they attempted to investigate biological complexity by fracturing it into its essential constituents. The focus of many biologists as well as the value placed by society and governments on biological research abruptly shifted from the organism to the cell and its parts. The era of the gene and the macromolecule began. The last decades of the twentieth century saw tremendous advances in our understanding of genetic control, cellular functions, biochemical interactions, and regulations, as well as the initiation of a complete sequencing of the metazoan genome.

The field of ecology in the era of environmental science also emerged during the last half of the twentieth century. As with reductionist science, initial attempts to investigate ecological complexity focused on fracturing ecosystems (e.g., niche and trophic) and landscapes (corridors and patches) into their essential elements.

In the new century we have a better understanding of biology at both the lower (i.e., cellular and molecular) and the higher (i.e., ecosystem, landscape, and global) levels of organization. We are poised to reassemble into a synthetic biological construct the elemental parts that have been carefully dissected over the last fifty years by the molecular and cell biologists at one end of the biological hierarchy as well as the ecosystem and landscape ecologists at the other end. Our next step is to develop a transdisciplinary science that integrates concepts, theories, and approaches across all levels of organization.

The consensual marriage of organismal biology and the advanced scientific tools of technology will be essential for our progress in understanding global "biocomplexity." Most scientists agree that global environments face a tremendous threat as human populations expand and natural resources are consumed. As natural habitats rapidly disappear the present century will be our last opportunity to understand fully the extent of our planet's biological complexity; this understanding will be dependent on the effective adaptation and utilization of new technologies. The contributors to this volume present their visions of biology in the next one hundred years with a spirit of biological discovery, with the foresight gained from in-depth knowledge of past achievements, and with an optimism grounded in reality.

These chapters are derived from invited lectures delivered at an international symposium (Biology: Challenges for the New Millennium) convened by the American Institute of Biological Sciences and the Smithsonian Institution from

March 22 through 24, 2000, at the Smithsonian's National Museum of Natural History in Washington, D.C. Building from those lectures, the scholars here attempt to outline the challenges that must be addressed in order to advance our science as humankind enters the twentieth-first century. The purpose of this volume is to examine the accomplishments of the past century and to provoke a thoughtful dialogue regarding the future challenges and opportunities in the field.

It is clear from the chapters in this book that a major challenge for biologists in the new century is to define an effective strategy for integrating the biological sciences with global economics and human social structure. The impending changes in our planet's basic biological functioning resulting from an unprecedented level of social and economic development will continue to have profound effects on human populations. Global well-being will depend on a concerted effort to integrate biological information, economic needs, and social organization. Each of the scholars who has contributed to this volume provides his or her own unique perspective on how this challenge will be addressed in the next hundred years.

Acknowledgments

We offer our deepest thanks to all the individuals and organizations that contributed to organizing the symposium "Biology: Challenges for the New Millennium" and publishing its papers in this book.

Preparations for the symposium began at the November 22, 1997, meeting of the Board of Directors of the American Institute of Biological Sciences (AIBS), when then President Elect Gary W. Barrett was appointed to chair a committee to organize the annual meeting for 2000. He suggested that internationally recognized scholars be invited to reflect on the accomplishments of the biological sciences during the past century and envision the opportunities and challenges of the twenty-first century.

Barrett felt that the "Year 2000" meeting (the institute's fifty-first annual meeting) should be held in Washington, D.C., which has been the headquarters of the AIBS since its founding in 1947, and that the Smithsonian Institution should be asked to cosponsor and host the event. On March 17, 1998, he met with Robert Hoffmann—a longtime friend and colleague—who introduced him to Barbara Schneider, senior program adviser in the Office of the Provost at the Smithsonian, to discuss the prospects for this meeting. Barrett submitted a letter to Smithsonian Provost J. Dennis O'Connor on April 23, 1998, to gain approval of these plans.

Hoffmann agreed to serve as cochair, with Barrett, of a twelve-member planning committee. Members representing the AIBS were Frank Harris, Kent Holsinger, Marilynn Maury, Richard O'Grady, and Hilary Swain; members representing the Smithsonian were Kathleen Desmond, John Kress, Michael Robinson, Barbara Schneider, and Dennis Whigham. We will always be grateful for the time, effort, and ideas offered by the Planning Committee, not only in helping to organize the meeting but in providing an infrastructure for this book. At its first meeting, on November 20, 1998, in the Smithsonian Castle, the committee agreed that "Biology: Challenges for the New Millennium" would be held from March 22 through March 24, 2000.

Next the Planning Committee, in consultation with the AIBS Executive Committee and additional scholars at the Smithsonian, drew up a short list of scientists to present plenary addresses and be honored for their contributions to biology. The process-oriented areas on which they were invited to comment included behavior, conservation, development, diversity, dynamics, energetics, evolution, integration, and regulation.

Once a consensus was reached on the meeting format, the AIBS Executive Committee, the Planning Committee, and the director of the Smithsonian Press, Peter F. Cannell, agreed to publish these presentations. W. John Kress and Gary Barrett agreed to serve as coeditors, and the speakers consented to publication of their presentations.

We sincerely thank these distinguished scholars for making this book possible. We are deeply indebted to Thomas Lovejoy for preparing an additional chapter. We extend special thanks to the following reviewers of these chapters (in alphabetical order): Brian Boom, New York Botanical Garden; Jane Brockmann, University of Florida; Paula DePriest, Smithsonian Institution; Patricia Gensel, University of North Carolina; Patricia Gowaty, University of Georgia; Gary Hartshorn, Duke University; Linda Kohn, University of Toronto; Gary Krupnick, Smithsonian Institution; Richard Norgaard, Berkeley, California; Lynne Parenti, Smithsonian Institution; Hilary Swain, Archbold Biological Station; Amy Ward, University of Alabama; and Judith Weis, Rutgers University. The assistance of Julie Barcelona in final editing is also much appreciated.

Finally, we thank Peter Cannell for his encouragement and helpful suggestions in bringing our efforts to fruition. We dedicate this volume to him in great appreciation of his love of books and his friendship.

Contributors

GARY W. BARRETT

Now Odum Professor of Ecology in the Institute of Ecology at the University of Georgia, Dr. Gary Barrett served as Director from 1994 through 1996. He was Distinguished Professor of Ecology at Miami University, Oxford, Ohio, until 1994; there he founded the Institute of Environmental Sciences and Ecology Research Center. He was Director of the Ecology Program at the National Science Foundation from 1981 to 1983 and has served on or chaired numerous committees in scientific organizations and professional societies, including the Ecological Society of America, the American Society of Mammalogists, the International Association for Ecology, and the National Research Council of the National Academy of Sciences. Dr. Barrett has served as President of the U.S. Section of the International Association for Landscape Ecology, the Association for Ecosystem Research Centers, and the American Institute of Biological Sciences, where he received the Presidential Citation Award in 2000. He has authored numerous books and scientific papers on ecology. Dr. Barrett received his B.S. from Oakland City University, his M.S. from Marquette University, and his Ph.D. from the University of Georgia.

DANIEL H. JANZEN

Dr. Daniel Janzen has devoted his scientific career to understanding and conserving tropical forests. He received his B.S. in botany and entomology from the University of Minnesota and his Ph.D. from the University of California at Berkeley. Previously on the faculties of the Universities of Michigan, Chicago, and Kansas, he is now Professor of Biology at the University of Pennsylvania. The author of numerous scientific papers and books, he has also served in an editorial capacity for four professional journals. He has been President of the Kansas Entomological Society, Counselor to the Society for the Study of Evolution, and an Executive Committee Member of the Organization for Tropical Studies. Among his awards are the Crafoord Prize in Coevolutionary Biology from the Royal Swedish Academy of Sciences and the Kyoto Prize in Basic Sciences from the Inamori Foundation. Dr. Janzen is a Technical Adviser for Conservation in Costa Rica, was made an honorary member of the National Park Service in that country, and has received the Award for the Improvement of Costa Rican Quality of Life.

W. JOHN KRESS

Dr. John Kress received his undergraduate training in biology from Harvard University and completed his Ph.D. at Duke University in 1981. He is Research Scientist, Curator, and Head of Botany at the National Museum of Natural History, Smithsonian Institution, and an Adjunct Professor of Biology at Duke University. He has also served as Honorary Curator at Fairchild Tropical Garden, the National Tropical Botanical Garden, and the Marie Selby Botanical Gardens. Dr. Kress is past President and current Executive Director of the Association for Tropical Biology and founder of the Tropical Biology Section of the Botanical Society of America. Before coming to the Smithsonian he was Director of Research at the Marie Selby Botanical Gardens. Dr. Kress has conducted research in both the New World and Old World tropics, concentrating on field studies of the systematics, phylogeny, reproductive biology, and conservation of large monocots, and has published numerous scientific and popular papers on tropical biology.

GENE E. LIKENS

Internationally recognized for his discovery of acid rain in North America and as cofounder of the Hubbard Brook Ecosystem Study, Dr. Gene Likens earned his B.S. from Manchester College (1957), his M.S. (1959) and Ph.D. (1962) from the University of Wisconsin at Madison, and eight honorary doctoral degrees for his contributions to ecology. In 1983 he joined The New York Botanical Garden, where he founded the Institute of Ecosystem Studies (IES). Since 1993 the IES has been an independent, not-for-profit research and education institution at which Dr. Likens is Director, President, and G. Evelyn Hutchinson Chair in Ecology. Dr. Likens's present university affiliations include Yale, Cornell, and Rutgers. He is a member of the National Academy of Sciences, the American Academy of Arts and Sciences, the British Ecological Society, the American Association for the Advancement of Science, and the Royal Danish Academy of Sciences and Letters. Among many awards he has received an American Institute of Biological Sciences Lifetime Accomplishment Award (2000), the Naumann-Thienemann Medal (1995), the Eminent Ecologist Award of the Ecological Society of America (1995), the Australia Prize for Science and Technology (1994), and the Tyler Prize for Environmental Achievement (1993).

THOMAS E. LOVEJOY

An important figure in tropical ecology and conservation biology, Dr. Thomas Lovejoy was instrumental in bringing the problems of tropical forest degradation to the public's consciousness. He originated the term *biological diversity* in 1980, initiated the early debt-for-nature swaps, founded the Minimum Critical Size of Ecosystem project in Amazonian Brazil, and was responsible for the public television series *Nature*. He has served as Executive Vice President at the World Wildlife Fund—U.S., Chief Biodiversity Adviser for the World Bank, and Science Adviser for the U.S. Secretary of the Interior and the Executive Director of the United Nations Environment Programme. Dr. Lovejoy was the first environmentalist to receive the Order of Rio Branco and the Grand Cross of the Order of Scientific Merit from the Brazilian government. He is past President of the American Institute of Biological Sciences and the Society for Conservation Biology, and past Chairman of the U.S. Man and Biosphere Program. He has authored numerous scientific articles and books on tropical biology and conservation. Dr. Lovejoy obtained his B.S. and Ph.D. degrees from Yale University. He is currently Counselor to the Secretary for Biodiversity and Environmental Affairs for the Smithsonian Institution.

LYNN MARGULIS

Dr. Lynn Margulis is a leading evolutionary biologist and contributed one of the most important explanations of the origin and diversification of life through her theory of serial endosymbiosis. Since 1988 Dr. Margulis has been Distinguished

University Professor in the Department of Geosciences at the University of Massachusetts at Amherst. She chaired the National Academy of Sciences's Space Science Board Committee on Planetary Biology and Chemical Evolution (1977–80), serves on the Science Council of NASA's Institute for Advanced Concepts, and codirects NASA's Planetary Biology Internship Program. An elected member of the National Academy of Sciences, the World Academy of Arts and Sciences, the American Academy of Arts and Sciences, and the Russian Academy of Natural Sciences, Dr. Margulis was awarded the National Medal of Science and Sigma Xi's distinguished Proctor Prize. Her books include *Symbiosis in Cell Evolution*, *Five Kingdoms* (with Karlene V. Schwartz), and *What Is Life?* (with Dorion Sagan). She holds ten honorary doctorate degrees, both national and international, in addition to her A.B. from the University of Chicago, M.S. from the University of Wisconsin, and Ph.D. from the University of California at Berkeley.

ERNST MAYR

One of the central figures in the New Synthesis of Evolutionary Biology, Professor Ernst Mayr has contributed immensely to our knowledge of speciation, systematics, ornithology, and the philosophy of science. He graduated from the University of Berlin in 1926 and became Assistant Curator in the Zoology Museum there. He joined the staff of the American Museum of Natural History in 1931 and served as Curator until 1953, when he was appointed Alexander Agassiz Professor of Zoology at Harvard University's Museum of Comparative Zoology. Professor Mayr was Editor and President of the Society for the Study of Evolution as well as President of the Society of Systematic Zoology. He is also a member of the National Academy of Sciences, the American Philosophical Society, the American Academy of Arts and Sciences, and the American Society of Zoology. He has received the prestigious Balzan Prize, the International Prize for Biology, and the Crafoord Prize, as well as the Wallace-Darwin Medal, the National Medal of Science, the Molina Prize, the Linnean Medal, the Gregor Mendel Medal, and the Benjamin Franklin Medal, among many other awards.

GORDON H. ORIANS

Dr. Gordon Orians is Professor Emeritus at the University of Washington, where he has been Director of the Institute of Environmental Studies and Professor of Zoology since 1960. He received his B.S. from the University of Wisconsin at Madison and his Ph.D. from the University of California at Berkeley. Dr. Orians has served as President of the Organization for Tropical Studies and the Ecological Society of America, as a member of the Board of Directors of the World Wildlife Fund and the Board of Visitors of the Organization for Tropical Studies, and as Chairman of the Board on Environmental Studies and Toxicology of the National Research Council. He was elected to the National Academy of Sciences, the Royal Netherlands Academy of Arts and Sciences, and the American Academy of Arts and Sciences, and was given the American Institute of Biological Sciences Distinguished Service Award and the Ecology Society of America Eminent Ecologist Award. He was Editor in Chief of the *Northwest Environmental Journal* and has held editorial positions for several scientific journals. Dr. Orians has made contributions to behavioral ecology, avian community structure, and the interface between science and environmental policy.

SIR GHILLEAN T. PRANCE

Currently the Scientific Director of the Eden Project in Cornwall, Visiting Professor at Reading University in the United Kingdom, and McBryde Professor at the National Tropical Botanical Garden in Hawaii, Sir Ghillean Prance was Director of the Royal Botanic Gardens at Kew from 1988 to 1999. He began his career at The New York Botanical Garden in 1963; he set up and directed the garden's Institute for Economic Botany from 1981 to 1988 and was Senior Vice President for Science there before going to Kew. He has served as President of the American Association of Plant Taxonomists, the Association for Tropical Biology, the Linnean Society of London, and the Systematics Association. He has received the New York Botanical Garden Distinguished Service Award, the Henry Shaw Medal of the Missouri Botanical Garden, the Victoria Medal of Honor from the Royal Horticultural Society, the Patron's Medal from the Royal Geographical Society, the Fairchild Medal for Plant Exploration, and the International Cosmos Prize for his environmental work in Brazil. His research interests include sustainable development and rain-forest conservation, especially in Amazonian Brazil, and he is the author of numerous books and scientific papers. Sir Ghillean holds fifteen international honorary doctorate degrees. He obtained his Ph.D. from Worcestershire and Keble College in Oxford.

MARVALEE H. WAKE

Dr. Marvalee Wake is Chair of the Department of Integrative Biology at the University of California at Berkeley. She is President of the Society for Integrative and Comparative Biology, President of the International Union of Biological Sciences, and a cofounder of DIVERSITAS, an international program in biodiversity science. Dr. Wake received a John Simon Guggenheim Memorial Foundation Fellowship in 1988 and is a Fellow of the American Association for the Advancement of Science and the California Academy of Sciences. She has also been a member of various scientific advisory boards and committees, including the National Science Foundation's BIO Advisory Committee, the American Institute of Biological Sciences Board of Directors, and the DIVERSITAS Executive Committee. Dr. Wake has authored numerous papers and books on evolutionary morphology and biodiversity science. An evolutionary morphologist by training and an integrative biologist by persuasion, she received her B.A., M.S., and Ph.D. degrees in zoology from the University of Southern California.

EDWARD O. WILSON

Professor Edward Wilson received his B.S. and M.S. degrees in biology from the University of Alabama and his Ph.D. from Harvard, where he is University Research Professor and Honorary Curator in Entomology at the Museum of Comparative Zoology. He is the author of two Pulitzer Prize–winning books as well as numerous other volumes and papers on evolution, behavior, and biodiversity conservation. The recipient of many fellowships, honors, and awards, Dr. Wilson has received the 1977 National Medal of Science, the Crafoord Prize from the Royal Swedish Academy of Sciences (1990), the International Prize for Biology from Japan (1993), the Gold Medal of the Worldwide Fund for Nature (1990), and the Audubon Medal of the National Audubon Society (1995), the last two for his efforts in conservation. He serves on the Boards of Directors of The Nature Conservancy, Conservation International, and the American Museum of Natural History.

A New Century of Biology

1

Introduction

The New Revolution in Biology

GARY W. BARRETT AND W. JOHN KRESS

As early as the 1970s numerous reports and publications stressed the crucial leadership role in social and economic development that biology would be required to assume as scientists began to plan for the end of the twentieth century and beyond. For example, Philip Handler (1970), then president of the National Academy of Sciences, noted that "the forces shaping the short-term future of man, perhaps to the turn of the Century, are apparent and the events are in train. The shape of the world in the year 2000 and man's place therein will be determined by the manner in which organized humanity confronts several major issues. If sufficiently successful, and mankind escapes the dark abysses of its own making, then truly the future belongs to man, the only product of biological evolution capable of controlling its own further destiny."

Further, a National Research Council report in 1970 from the Committee on Research in the Life Sciences (NRC 1970), chaired by Harvey Brooks, Harvard University, listed ten frontiers of biology, ranging from the origin and language of life to the diversity of life. However, this report also stressed the deterioration in quality of the planet's air, water, and soil fertility. Critics noted that this deterioration was proceeding most rapidly in technologically developed nations and that this growing threat to the quality of life and, indeed, to the habitability of the planet constitutes a profound human issue. This report noted that "although the life sciences, even now, are capable of contributing significantly to this critical enterprise, the science of ecology, while crucial, is still developing; its capabilities are limited as is the number of ecologists. It must be clear that ecological understanding rests upon the totality of all other biological understanding. . . . Thus, continuing advancement of understanding along all biological fronts is essential to the development of ecological understanding." Here we should note that the 1970s were frequently referred to as the "Decade of the Environment." Needless to say, the marriage of biological progress with ecological understanding continued through the remainder of the twentieth century.

In the proceedings following an international conference entitled "Biology and the Future of Man," held in Paris in 1974 (Galperine 1976), Valéry Giscard d'Estaing, president of France, stated, "There is no doubt that mathematics, physics, and other sciences rather ill-advisedly referred to as 'exact'—very likely by contrast with economics—will continue to afford surprising discoveries. Yet, I cannot help feeling that the real scientific revolution of the future must come from biology."

Thus, since at least the 1970s an optimism has prevailed that biology would play a lead role in improving the quality of humankind, including the environment in which we reside. The chapters in the present volume illustrate that we not only have avoided most of the abysses Handler referred to but are rapidly developing the technologies, database, and transdisciplinary approaches necessary to solve the problems that loom in our future. As societies begin to address challenges and opportunities associated with a world at carrying capacity (Barrett and Odum 2000), we visualize a future with a "glass more than half full" thanks to accomplishments during the past century. The authors of the following chapters provide a glimpse of the twenty-first century and how the life and environmental sciences are likely to seed a new revolution in integrative biology as a result of this era of accomplishment.

THE NEW BIOLOGY

Without question the twentieth century was a seminal epoch for the study of biology as measured by its conceptual advances, its research support for areas such as medicine and agriculture, and its growing importance to our future (Bock 1998). At the close of the twentieth century numerous scholars and reporters attempted to highlight events and milestones of the past hundred years in fields such as military battles, climatic phenomena, athletics, and advanced technologies. The encyclopedic accomplishments and discoveries in biology, however, were so diverse and profound that few attempts have been made to provide a comprehensive overview. Instead summaries or reflections were presented in select areas, such as the top one hundred or so books that shaped the century of science (Morrison and Morrison 1999). Likewise, several scholars have attempted a vision regarding how science will revolutionize the twenty-first century (Murphy and O'Neill 1995; Kaku 1997). However, for the most part large professional societies did not simultaneously try to reflect on major accomplishments of the past century *and* to visualize the challenges and opportunities that biologists must address during the next. Such a synthesis is one of the major goals of this book.

At the beginning of this new century we are faced with a biological paradox of the last one hundred years—namely that the unparalleled advances in areas such as molecular, cellular, and organismal biology were coupled with greatly increased problems related to the world's human population growth, food production, energy resource management, biotic diversity, and landscape fragmentation. Ernst Mayr in his book *This Is Biology* (1997), an excellent overview of the field during the past century, addressed this paradox. He not only stressed the integrative role of evolutionary theory during the last hundred years but pointed out that biology at the organismal level and below will need to be coupled with holistic approaches as societies address problems related to overpopulation, environmental degradation, and resource management. He wrote, "Overpopulation, the destruction of the environment, and the malaise of the inner cities cannot be solved by technological advances, nor by literature and history, but ultimately only by measures that are based on an understanding of the biological roots of these problems." It is no wonder that Ernst Mayr, a leading evolutionary biologist of this century, provides the foreword to the present volume.

The core chapters of this volume are devoted to contributions from internationally renowned scientists who were invited to address the processes (i.e., energetics, evolution, development, regulation, behavior, diversity, and conservation) that transcend all levels of biological organization (see Figure 1.1; Barrett et al. 1997). The authors were selected to maintain (a) a balance among the levels of biological organization; (b) an emphasis on integrative processes among levels of organization; and (c) an organization that highlighted transdisciplinary accomplishments in all fields of the life sciences rather than narrow intradisciplinary work. These chapters thus represent a breadth of insight and experience across a wide range of the biological sciences.

CONCEPTS AND OPPORTUNITIES IN THE TWENTY-FIRST CENTURY

Here we outline ten integrative topics that provide food-for-thought concepts for the volume as a whole. These areas are not intended to be in any ranked order nor are they all-encompassing. We hope that they will provide readers, especially graduate students and others wanting to better understand opportuni-

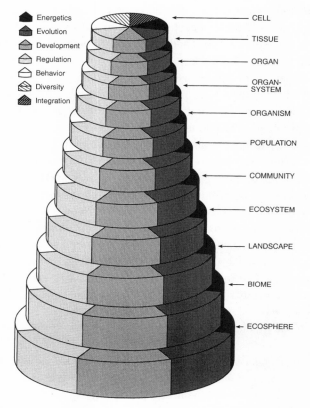

Figure 1.1. Model of transverse processes across levels of biological organization (from Barrett ~* al. 1997).

ties in biology, with challenges that will contribute to making the twenty-first century one that will be recognized for the integration of humankind with our biological and technological systems.

- *Synergism.* Throughout each chapter readers are encouraged to understand how synergistic and mutualistic processes operate within cells, organisms, and populations at one level as well as between ecological and social systems at another. *Synergy* is defined here as combined action or operation. Lynn Margulis (Chapter 2) attacks the neo-Darwinian paradigm and presents new ideas on the interactions of microbes in the origins of species, especially with regard to the source of evolutionary novelty. She describes the evolution of complex unions in the origin of the eukaryotes and outlines the microbial contribution to the evolution of life.

- *Optimization*. Several chapters portray how evolutionary processes, including morphologies, behaviors, and energy transformations, are optimized for functional efficiency at each level of organization. This optimization process ranges from transfer of genetic information at the molecular level to energy efficiencies at the ecosystem level. More specifically, Marvalee Wake (Chapter 3) addresses the optimization and integration of morphology (as the structural basis for an organism's interaction with the environment) and development (as the process by which morphology is achieved). She suggests that the monogamous marriage of developmental studies with evolutionary theory is clearly the foundation of a new, if not rejuvenated, discipline in the biological sciences. She also speculates on the future polygamous relationship of developmental biology with evolutionary, ecological, genetic, and molecular sciences. In Chapter 5 Gordon Orians asks the question "How does animal behavior intersect with modern concepts of evolutionary and ecological theory?" He recognizes that it is hard to admit that the concept of free will may be compromised by an evolutionary explanation of human behavior. He also points out that parsimonious interpretations of phylogenetic data are powerful new tools in the study of the evolution of behavior.

- *Optimism*. Without exception contributors note that the twenty-first century will result in less emphasis on disciplinary training and greater focus on interdisciplinary and team approaches to research and problem solving. Gene Likens (Chapter 4) points out that interactions between natural ecosystems and society must be a major concentration of biologists in the near future. He stresses the practical aspects of scientific team building, the evaluation of ecological complexity, the accumulation of long-term environmental data, and the use of new technological tools as priorities and opportunities for solving ecological issues in the future.

- *Conservation*. Conservation—the careful preservation or protection of an entity—is a common theme throughout this book. Whether one is referring to the conservation of energy, matter, genetic material, endangered species, or wilderness areas, this priority will increasingly dominate resource management, political decision making, and national and international research cooperation well into the twenty-first century. Society increasingly views Biosphere I—the Earth—as a total system intricately interconnected and possessing regulatory mechanisms and mutualistic relationships. These processes—the core of these chapters—and relationships must be preserved to ensure a quality of life for all. A very clear picture of our biologically stressed planet is provided by Sir Ghillean Prance (Chapter 6) and Thomas Lovejoy (Chapter 7), who note that there is still much to learn about the diversity of life. Over twenty thousand new species of vascular plants alone, one of the best known groups of organisms, were described over a recent nine-year period, indicating that systematists have much work remaining in the inventory of the world's taxa. Prance also bluntly challenges biologists to accept the responsibility of addressing the political nature of what they do as natural historians.

- *Sustainability*. Robert Goodland (1995) defines sustainability as "preserving nature's capital." By understanding the processes (e.g., evolution, regulation, and integration) that transcend all levels of organization, societies will be informed in a way that programs such as preserving biotic diversity will be based on an educational incentive rather than a regulatory mandate. Such a research, educational, and service agenda should result in both sustainable ecological systems and a sustainable society. Daniel Janzen develops this central theme in his discussion of the "gardenification of nature" (Chapter 8). After years of effort in Costa Rica, he reluctantly accepts that we will never have a smooth transition between society and the preservation of biodiversity, but he insists that we must "know it (i.e., biodiversity) and use it in order to save it."

- *Consilience*. Edward Wilson (Chapter 9) defines *consilience* as cause-and-effect explanation across disciplines. Exploring consilience requires experiments and research designs resulting in strong inference (Platt 1964). These designs and approaches, whether they are established at or within molecular or human-dominated systems, should continue to revolutionize the biological sciences well into the twenty-first century. Such investigations will require a level of cooperation and unity of knowledge based not on technologies (which are largely available at present) but on human values and societal enlightenment. Wilson suggests that the great future frontiers of biology will include evolutionary genomics, biodiversity research, large-scale community ecology, and linkage of biology to the humanities and the social sciences. He views human nature as the product of the epigenetic rules of our evolutionary heritage, which can only be understood within the perspective of our biological nature.

- *Holism*. During the past two decades several scholars (e.g., Brown 1981) called attention to two "schools of research philosophy," namely the ecosystem ecology school, led by Eugene P. Odum, which concentrated on the flow of energy and matter through systems and emphasized mutualistic interactions between organisms and the physical environment, and the evolutionary biology school, led by the late Robert MacArthur, which concentrated on evolutionary interactions between and among species with emphasis on community organization, competition models, and species diversity. It is apparent, as we enter the twenty-first century, that systems ecologists are including evolutionary factors (e.g., secondary chemistry, microbial interactions, and foraging strategies) in systems models, and that evolutionary biologists are increasingly addressing macroecological factors (e.g., global climate change, landscape patch diversity, and nutrient cycles) as they investigate questions at the microevolutionary scale. In other words, as we enter this new revolution in biological thought, there exists a coupling of reductionist science and holistic science, including the social sciences. Barrett and Odum (1998) refer to this holistic and integrative approach to addressing societal problems as "integrative science"—a new type of science developed along the lines of C. P. Snow's "third culture" (Snow 1963).

In summary, we trust that *A New Century of Biology* will give cause to reflect and to appreciate the evolution of the biological sciences. We also hope it will give cause for all to participate in this new biology during the twenty-first century. It provides a niche for each citizen of every society in the years ahead.

LITERATURE CITED

Barrett, G. W., and E. P. Odum. 1998. From the president: Integrative science. *Bio-Science* 48:980.

——. 2000. The twenty-first century: The world at carrying capacity. *BioScience* 50:363–68.

Barrett, G. W., J. D. Peles, and E. P. Odum. 1997. Transcending processes and the levels-of-organization concept. *BioScience* 47:531–35.

Bock, W. J. 1998. The preeminent value of evolutionary insight in biological science. *Amer. Sci.* 86:186–87.

Brown, J. H. 1981. Two decades of homage to Santa Rosalia: Towards a general theory of diversity. *Amer. Zool.* 21:877–88.

Galperine, C., ed. 1976. *Proceedings of the international conference Biology and the Future of Man.* New York: McGraw-Hill.

Goodland, R. 1995. The concept of environmental sustainability. *Ann. Rev. Ecol. Syst.* 26:1–24.

Handler, P., ed. 1970. *Biology and the future of man.* New York: Oxford Univ. Press.

Kaku, M. 1997. *Visions: How science will revolutionize the twenty-first century.* New York: Bantam Doubleday Dell.

Mayr, E. 1997. *This is biology: The science of the living world*. Cambridge, Mass.: Harvard Univ. Press, Belknap Press.

Morrison, P., and P. Morrison. 1999. One hundred or so books that shaped a century of science. *Amer. Sci.* 87:542–53.

Murphy, M. P., and L. A. J. O'Neill. 1995. *What is life—The next fifty years: Speculations on the future of biology*. Cambridge: Cambridge Univ. Press.

National Research Council. 1970. *The life sciences: Recent progress and application to human affairs*. Washington, D.C.: National Academy of Sciences.

Platt, J. R. 1964. Strong inference. *Science* 146:347–53.

Snow, C. P. 1963. *The two cultures: A second look*. New York: Cambridge Univ. Press.

2

Bacteria in the Origins of Species
Demise of the Neo-Darwinian Paradigm
LYNN MARGULIS

THE LAST CENTURY

Evolution, the science of the changes in live organisms and their populations through time, provides the organizing principle for all the biological disciplines. "Nothing in biology makes sense," wrote Theodosius Dobzhansky (1973), "except in the light of evolution." In one of the most important scientific contributions in the history of human knowledge, Charles Darwin (1809–82) posited that all life on Earth is related through "descent with modification." Although he did not use the term *evolution*, he developed the idea: more live beings are produced, hatched, born, et cetera than can possibly survive. Variation is observable in all populations of live beings at all times. Some of this variation is inherited by descendants so that populations of organisms change gradually through time. New species descendants come into being from their ancestors. This process has an unbroken history observable in the geological record as a succession of fossils relatable to extant forms of life.

These tenets of Darwinism are as fresh today as they were at the end of the nineteenth century. In the early twentieth century attempts to fuse Darwin's ideas of gradual biological change with Gregor Mendel's (1822–84) concepts of the stasis of hereditary factors (later termed *genes*) led to a body of literature called the new synthesis or the neo-Darwinian paradigm. Population genetics was married to explanations of changes through time primarily of diploid, outbreeding terrestrial animals, and *evolution* was redefined as "changes in gene frequencies in natural populations through time." Both the fossil record and the history of the environment were ignored—the details of these observations passed to the geological (long-term) and ecological (short-term) sciences. Evolutionary change was attributed primarily to immigration, emigration, random mutation, breeding structure, founder population size, and chromosomal rearrangements (karyotypic changes). Critics of neo-Darwinian evolution were silenced, ridiculed, ignored, or denied research funds. In this new century the critics will flourish and the neo-Darwinist demise will be complete.

Biologists of the new century will realize that microbes, live beings too small to be seen without the aid of microscopes, provide the mysterious creative force in the evolutionary process. Bacterial acquisition and integration is the most important factor in the origin of species. Prokaryotes, organisms with bacterial cell structure, are champions at passing genes and other information. The machinations of bacteria and other microbes, not just frequency changes, underlie the whole story of Darwinian evolution.

Free-living bacteria and other microbes tend to merge with larger forms of life, sometimes seasonally and occasionally, sometimes permanently. Inheritance of "acquired microbes" often ensues under conditions of unmitigated stress. With Dobzhansky many have noted that the complexity and responsiveness of life, including the appearance of new species from differing ancestors, can be comprehended only in the light of evolutionary history. But in this chapter I will show that today's evolutionary saga itself, legitimately vulnerable to criticism from within and beyond science, is entirely incomprehensible if bacteria are omitted from the story. I submit that if Charles Darwin, who sought the source of new inherited variation and the origins of new species, only knew about the subvisible world what we know today, he would have chuckled. He would have laughed out loud. Then he would have agreed with me.

FOLK MEDICINE, FOLK GENETICS

For most of human history physicians, shamans, and herbalists practiced the healing art in a state of well-educated ignorance. No science of genetics existed, even though astute practitioners recognized that certain diseases prevailed in certain families. Among Hebrews in biblical times, for instance, it was considered ill-advised to circumcise any boy whose uncle, but only on his mother's side, had suffered excessive bleeding at his own circumcision. To predict a woman's baldness or her baby's "cri du chat" (Tay-Sachs) disease, doctors traced the trait on both sides of the family. But practical genetic advice, culturally transmitted as learned folklore, never stemmed from a logical body of knowledge. Until the beginning of the twentieth century, nothing was known of the causes of inherited diseases—either their physiology or the reasons they ran in families—because no one, physician or not, knew anything about transmission of genes or chemistry of DNA and other nucleic acids.

CATCHING GERMS

Conscientious medical men and midwives suffered enough plague, childbed fever, pox, and other "virulence" to grasp firmly the concept of contagion. Yet a coherent philosophy of contagious disease was elusive, and hospitals remained excellent final resting places throughout most of European history. Neither health professionals nor the general public was aware of the existence of the microcosm, the world of the very small, until the work of Louis Pasteur and Robert Koch on the germ theory of disease. Only after Pasteur made palpable the habits of bacteria and yeast, and Koch developed his famous postulates for the proof that "germs cause disease" did the subvisible microbial presence begin to insinuate into our cultural heritage. By the first half of the twentieth century, thanks in part to Paul de Kruif's *Microbe hunters* (1953), young mothers and octogenarian physicians alike were utterly convinced that bacteria, or "germs," must be eradicated from their lives.

Germs, like *weeds* or *toadstools,* was an undefined but clearly understood term: it denoted an unwanted form of life. Of the subtle stages of the malarial parasite (*Plasmodium*) or nutritional interchange between *E. coli* and its human intestinal habitat, the infant science microbiology still contributed little. Physicians did not much more than classify by symptom and prescribe safe treatment: Smoke, herbal infusion, aspirin, morphine, cocaine, extirpation, and many other "remedies" filled the black bag of tricks. Desperately ill patients were treated by "cupping," compresses, poultices, chanting, and "laying on of hands." The major antidote, as from time immemorial, was confident reassurance dispensed by specialists at high price.

Evolutionary theory begins the new century much the way medicine did a hundred years ago. The search for new knowledge is inevitably embedded in culture. Social practice has hindered even professional evolutionary theorists, who tend to be abysmally ignorant of the science most relevant to their work. The situation in contemporary evolution may even be worse than that in nineteenth-century medicine. The facts required by turn-of-the-century doctors did not yet exist, whereas the science background essential to evolutionists exists but is systematically ignored.

ACADEMIC APARTHEID

Ironically, science has documented evolution in action even if individual scientists have not. Most information relevant to understanding evolution lies hidden in arcane literature. The news does not reach the professionals or the public. A fragmented body of literature, detailed but disorganized, tells how species originate and how the complex novelty of living beings appears and spreads. Mainly recorded in the esoteric languages of biochemistry and microbiology, it is inaccessible even to professional biologists and geologists who specialize in evolution. Biochemists, microbiologists, physiologists, and other experimenters in esoterica often avoid discussion of the evolutionary implications of their work; they disdain them as "speculation." They prefer to tackle questions answerable by direct evidence rather than "philosophy." Many regard the scientific reconstruction of prehistory, a subject replete with inference and judgment, as a dirty practice.

Isolation of scientists in fields relevant to evolution is exacerbated by terminological differences. They lack ways to integrate the myriad idiosyncratic descriptions of fundamental discoveries in live organisms into thought patterns of their artificially bounded "fields." Dorion Sagan and I are in the throes of writing a book to be entitled *Origins of species: Inheritance of acquired genomes* (Margulis and Sagan, in press), which will connect experimental with theoretical modes of explanation. We will be able to leap the communication barrier because we stand on the shoulders of twentieth-century and earlier contributors to the science of biology. We plan to reveal what evolutionary science really knows about how species begin in a manner accessible to nonspecialist readers.

DARWIN AND LAMARCK

Charles Darwin established to the satisfaction of his scientific contemporaries and followers that species descend from their predecessors. All life is connected back through time to preexisting forms, ultimately to the origin of life itself. Darwin described how living things "beget" descendants that differ slightly from their parents. He showed that many of these differences are inherited. Of the large numbers of offspring that potentially grow from seed, hatch, or are born, only a limited population actually survive. These survivors, by his logic, must have traits that are more conducive to survival in that particular environment than those in the offspring that didn't survive.

Darwin gave this process of differential survival and reproduction the name *natural selection*. But *natural selection* denotes only the survival and reproduction of the few relative to the prodigious many. The process of natural selection, in spite of loud protest to the contrary, does not by itself create novelty. *Natural selection,* really differential survival, only selects from traits that already exist. So how did intrinsic variation first arise? Darwin, like his predecessor Jean-Baptiste Lamarck (1744–1829), struggled with the problem. Darwin invented, in fact, a Lamarckian explanation. He proposed the "pangenesis" hypothesis for "the inheritance of acquired characteristics." He tried to explain the source of evolutionary innovation, but ultimately he simply admitted his ignorance.

THE SOURCE OF EVOLUTIONARY NOVELTY

The question of the genesis of naturally inherited variation has baffled many since Darwin. Zoologists tend to argue that random changes in DNA, primarily base pair mutations, have generated all of evolutionary change. Botanists are often able to correlate chromosome and chloroplast differences inside plant cells with the appearance of new horticultural species. Although plastid-genome relations and chromosome changes relative to hybridization lead to new variation—even species—botanists cannot generalize their explanation to algae, animals, or microbes. Biochemists and physiologists simply have no idea how the complexity of life ever evolved; in any case, they do not write about it. Most ignore the whole of evolutionary science.

The phenomenon of evolution occurs over the entire face of the Earth, from at least 12 kilometers below sea level, in the ocean's deepest abysses, to as many as 8 kilometers above sea level on the highest mountains. Recently life has been recorded in wet fissures of granitic rocks 3 kilometers into the Earth's lithosphere and at hot-water vents on the sea bottom. Life began at least 3.8 billion years ago. The detailed record of evolution, preserved in rocks both as fossils and as short- and long-chain extractable carbon compounds, overwhelms those who study it. Bacteria and other cells reflect their evolutionary history. Yet in spite of the diversity of clues by which the evolution of life is reconstructible, most self-described evolutionary biologists eschew cell biology, microbiology, and the geological rock record. They are zoologists, concerned only with animals. Nothing prior to the last 541 million years enters their line of sight. The crucial science—aeons, geographies, microbiology, protistology, and cell chemistry—that holds the answer to Darwin's enigma is, in the end, formally eliminated from consideration by zoological evolutionary biologists, who define the boundaries of the relevant.

TWENTY-FIRST CENTURY OUTLOOK

I reiterate the value of Darwin's original all-inclusive stance. As Ernst Mayr, Stephen Jay Gould, and other superbly educated scientists have always insisted, evolutionary concepts are necessarily multidimensional and interconnected. They must provide organizing principles for understanding life. Here I mention the development of four ideas, which we plan to elaborate on in each section of our forthcoming book (Margulis and Sagan, in press). I here outline the science already known about how species really begin and proffer some reasons good science is so often ignored. These are the four concepts:

1. *Evolution always works because of the tendency for exponential growth of populations that must be checked* (i.e., growth coupled with the inhibitory nature of natural selection).
2. *Most biologists do not really appreciate the astounding metabolic repertoire of bacteria.* That bacteria are not organized naturally into species is not understood either. Bacteria population growth, community organization, gas production, motility and sensitivity, propagule formation, and resistance are but a few of the crucial but ignored phenomena intimately associated with evolution of the rest of life.
3. *The largest discontinuity among extant life-forms on this planet is that between prokaryotes (bacteria) and eukaryotes (all the others).* The evolutionary history of this discontinuity includes the evolution of the process of speciation. No speciation was possible before the origin and evolution of eukaryotes by symbiogenesis.
4. *The extent of "hybridization" or metaspecies "fertilization" as theoretically forbidden (sexual and parasexual) mergers is astounding.* This bizarre but fertile genetic recombination between organisms not simply of complementary genders of the same species occurred over and over in the history of life. In some infrequent cases new higher taxa were generated. The sexual and parasexual fusions of members of different species have had important evolutionary consequences. Viable sexual and parasexual mergers when cyclically repeated led to new groups, including new species.

All these concepts illustrate from different vantages the same fundamental yet poorly known idea: The agents of evolutionary change are microbes, only supplemented, not supplanted, by random mutations. The formation and diversification of a new species is the outward manifestation of the actions of subvisible forms of life: bacteria, protists, and fungi. Evolution emerges from the fact that small living organisms, like all others studied by Darwin, tend to outgrow their bounds. The disguised or unseen beings that decimate our populations with virulent disease and provide soil nitrogen to our food plants also play the major creative role in the genesis of new life-forms.

MICROBIAL EVOLUTIONARY DIVERSITY

A prodigious technical literature shows that bacteria are the main repository of evolutionary diversity. The entire animal kingdom employs essentially one mode of metabolism: the use of oxygen to respire organic food molecules (heterotrophy). Plants employ two: heterotrophy via oxygen, just as in animals, and oxygenic photosynthesis. Bacteria have, in addition to these two, at least twenty fundamentally different metabolic modes (Balows et al. 1992). None is known in either animal or plant.

Some bacteria breathe sulfur or arsenic rather than oxygen. Others glow in the dark by complex luciferin-luciferase bioluminescent reactions. Some thrive in foul-smelling communities with swimmers that transform metals such as manganese or iron. Certain bluish or green ones photosynthesize and breathe oxygen the way plants do, whereas others use sunlight to photosynthesize in ways that preclude either release or breathing of oxygen—ever. They use hydrogen or hydrogen sulfide, but never water, as their electron source. Still different bacteria change carbon dioxide and hydrogen into swamp gas; that is, they make methane, the same gas we burn in stoves. The swamp gas flows into crevices, where still another obscure type of bacterium, which enjoys a metabolic mode called methylotrophy, burns methane to thrive. Fiercely predatory bacteria swim ten times their body length each second. They respond instantly to mechanical stimuli and kill choice bacterial victims with 100 percent efficiency. Still different bacteria grow by miniaturized sunlight-based chemical reactions at huge salt concentrations that clever scientists can neither explain nor imitate. Our cultural dismissal of bacteria as "primitive," "simple," and "eradicable" is misguided.

Bacteria and their protist descendants were capable of astounding feats well before any animals evolved. Certain protists are master engineers; others invented agriculture. Single-celled foraminifera, huge by one-cell standards, farm trapped algae inside their shell-covered bodies. The algae are expelled during daylight hours and returned to pens inside the shells at night. Some of their fussy relatives make new shells with care: They choose black pebbles from a multi-color mix and plaster the tiny stones to their bodies. Other foraminifera even pack and paste tiny particles together to build lookout towers. They climb atop the towers and hunt animals whose bodies are far larger than these protists. Fungi, no different from patches of tiny white scum to the unaided eye, wrap their stringy bodies to become spring-loaded traps. Like pythons, they squeeze victimized nematode worms to death. Microbes, in short, evolved to act in ways we associate only with familiar animals and plants.

BEYOND BOTH LAMARCK AND DARWIN

How was microbial creativity transferred to larger life-forms? A main suggestion for the new century of biology is that the maligned Lamarckian slogan "inheritance of acquired characteristics" should still not be abandoned: It just needs to be carefully refined. No, individual plants and animals do not acquire heritable traits by growing, eating, exercising, mating, and the like. Rather, under stress, different kinds of highly endowed individuals physically associate. Some may later incorporate and still later even fuse their genetic systems. Many fusion modes, including virus infection, are documented (Margulis 1993). The associates always leave behind free-living, unassociated relatives. Permanent mergers of bacteria, protists, and/or fungi with each other or with plant and animal "hosts" generate large-scale evolutionary change. Rapid acquisition of new, highly refined microbial traits confers selective advantage on their plant or animal captors. The inheritance of trapped microbial populations creates novel, unprecedented lineages. Random genetic mutations refine and hone, but never by themselves do they create inherited variation.

Just as marriage or corporate merger cannot be simply reversed, evolution by microbial acquisition over the long term becomes an irreversible process. New species of plants and animals evolve by "inheritance of acquired genomes," which, after integration into a partner's genome, no longer can be "unacquired."

I submit that knowledge of the microcosm is essential for the solution to Darwin's question "What is passed from parent to descendant that we detect as evolutionary novelty?" A straightforward answer is "Populations and even communities of microbes." Whereas "populations" are individuals of the same type living at the same time and place, "communities"—populations of different types living at the same time and place—prevail in nature. Communities may actually fuse and transfer their genes among their members. Thus new, large, more complex "individuals" evolve. All evolutionists and systems ecologists must learn microbiology.

The ways in which microbes tend to join each other physically and to interact both among themselves and with much larger associates have been told in unruly form in the specialized language of the sciences. Even those of us who understand how much is already known about the origin of species are limited to work on our own tiny discoveries, usually one species at a time. Academic "biology departments" splinter into "molecular" versus "organismal" biology, which exacerbates misunderstanding. The relevant information is scattered in more than a dozen esoteric languages: algology (phycology), bacteriology, biochemistry, cell biology, geology, invertebrate zoology, metabolism, microbial ecology, molecular evolution, nutrition, paleontology, protistology, sedimentary geology, and virology, among others. These fields, black boxes to the public, are mysteries even to many who practice "evolutionary biology" today.

I posit that all of neo-Darwinism will be considered with a smirk to have faded as a twentieth-century anglophone prejudice, aberration, and flawed mental exercise. Neo-Darwinism moved too far away from life—as H. Minakata, after whom the slime mold genus *Minakatella* is named, said, the English-American biologists tried to take the life out of biology (NHK-TV 1990). The new century will begin again where Darwin ended. Entirely consistent with Darwin's great insight, the new evolution will reveal science beyond his century. It will show how the luxuriant living diversity that surrounds us evolved in small but discontinuous steps. What appear as magic, "irreducible complexity," or "grand design"—the retinal image of the vertebrate eye, the bumblebee's flight, the ocean-traversing song of the humpback whale or of Luciano Pavarotti—are legacies of repeated interactions. Familiar organisms, inevitably large, and their motivations, protrude from the microbial underworld. The evolutionary powers of the microbial ancestors of us all deserve far wider recognition.

HOW DOES EVOLUTION WORK?

Darwin's insight into evolution led to scientific understanding of population growth, the existence of heritable change, and the force of natural selection. Nature is organized. Organisms live in communities made of populations. All communities comprise different species that live together in identifiable habitats. Such organization in nature precisely correlates with climatic, geographic, and other environmental factors. Taxonomy, the science of naming, identifying, and grouping live beings, tends to ignore environmental correlates and is, in the case of the bacteria, artificial. Even so, in protists, fungi, plants, and animals, the most easily identified and named group level is that of species. About 30 million extant, and perhaps a thousand times more extinct, species have been estimated. Species are distinguished, counted, documented, and named based on direct observation. Very few are mated to determine their existence. Although many types of bacteria are recognizable, we believe that bacteria populations are not organized into stable species in the way that nucleated organisms are. The species concept simply does not apply. Bacteriologists say that if two kinds of bacteria share 85 percent of their traits, they belong to the same species: 86 percent same species; 84 percent different species. This highly arbitrary measure differs greatly from any species delineation in eukaryotes. The process of speciation itself evolved only in the lower Proterozoic aeon, about 2.5 billion years ago, when the transformation from bacteria to larger forms of life occurred.

Every organism, with or without a mate, enjoys an intrinsic capacity to increase geometrically its number of offspring. Populations "unchecked" (Darwin's term) overgrow their limits. Daily environmental constraints, such as lack of water, crowding, and threat of starvation, prevent populations from the indefinite expansion of which they are capable. Since each population has specific energy, nutrient, water, and space requirements that are never fully provided by the environment, population expansion is inevitably stressed. Natural selection, a strictly subtractive process, eliminates all who fail to survive for any reason. Those who remain, by definition, survive and tend to pass on heritable traits to their offspring. Since survivors retain traits most conducive to survival at given times and places, life on Earth retains a memory of its past. Living bodies store in their complex chemistry memories of past environmental limitations they overcame.

Because it is nothing more than differential survival, natural selection perpetuates, but it cannot create. Then what generates evolutionary innovation? The list is now far longer than random mutations of individual genes. Novelty appears and accumulates not only by single base pair changes (random or not) in DNA but by duplication or movement of genes or blocks of genes and accumulation of viruses, plasmids, and other short pieces of DNA. Cells and organisms acquire long pieces of DNA by bacterial mating, chromosome rearrangement, symbiont acquisition, and hybridization. Cross-fertilization of plants and animals with members of the same or even of different species does occur. When these phenomena are taken together, a strong case is made that Darwin's dilemma is solved. We now know the sources of evolutionary novelty—I have space only to list them here.

THE MICROBES' CONTRIBUTION TO EVOLUTION

We will see in this new century the refutation of the microbes' notoriety. Their bad reputation will be replaced with the glorious description deserved by these ingenious life-forms. Evolution's inner workings require understanding of the full range of life's possibilities. We must work to explain the extent of bacterial evolutionary complexity. Bacteria reproduce without sex, and they duplicate, transfer, digest, and in other ways lose and gain genes. The speed, volume, and antiquity of bacterial gene-trading activities amaze us.

The major bacterial modes of nutrition are listed in Table 2.1. Much is known about them. Compared with those of plants, animals, and fungi, the metabolic repertoire of bacteria and their protist descendants is vast. Long before the appearance of animal or plant life, movement, photosynthesis, predation, sexuality, gender, immunity, and other attributes of the living were already exquisitely refined in the bacterial world. The metabolic virtuosities of bacteria and protists are organized into a wet microbial World Wide Web that preceded the human machinate one by at least 2 billion years. The environment within which evolution occurs is dynamically stable, self-regulating, and largely maintained by the chemical and biological interactions of members of microbial communities. Ironically the scientists who most need to know about this, ecologists and evolutionists, are among the least informed.

POWER TO THE PROTISTS

Permanent mergers of very different bacterial types led to the first anaerobic nucleated cells. Casual relationships that became irreversibly intimate generated the protists. Protists, smaller members of the great Protoctista kingdom, were the microbial ancestors to all large forms of life. A little-known but easily demonstrated fact is that some nucleated cells (also called eukaryotic, the kind that makes up the bodies of all protists, fungi, plants, and animals, including of course humans) live quite happily without any oxygen. Indeed, to them oxygen is an instant poison. Among them are the amitochondriate protists, a group of hundreds of species that live in habitats such as the intestines of mammals and insects or sulfur-rich muds (Margulis et al. 2000; Figure 2.1). Amitochondriate protists are living representatives of the ancestors of all nucleated organisms. They cannot metabolize oxygen nor can any reproduce sexually. However, these obscure microbes do show speciation.

Species originated, in my view, in eukaryotes before the appearance of either oxygen respiration or the kind of sexuality that involves fertilization and cell fusion. The familiar phenomenon of stable species that can be named, identified, and classified, which exists nowhere in the bacterial world, started with protists (the earliest protoctists). Unlike plants, animals, and other eukaryotes, bacteria show a continuum of traits such that "species" changes are easily induced. Bacterial "species" change on a daily, weekly, or monthly basis (Sonea and Mathieu 2000). The phenomenon of speciation itself is a product of evolu-

Table 2.1

Modes of Nutrition for Life on Earth (Names constructed by adding the suffix *-troph*, e.g., photolithoautotroph [plant])

Energy (Light or Chemical Compounds)	Electrons (or Hydrogen Donors)	Carbon Source	Organisms and Their Hydrogen or Electron Donors

PHOTO- (light)	LITHO- (inorganic compounds and C_1)	AUTO- (CO_2)	PROKARYOTES
			    Chlorobiaceae, H_2S, S
			    Chromariaceae, H_2S, S
			    Rhodospirillaceae, H_2
			    Cyanobacteria, H_2O
			    Chloroxybacteria, H_2O
			    PROTOCTISTA (algae), H_2O
			    PLANTS, H_2O

		HETERO- $(CH_2O)_N$	NONE
	ORGANO- (organic compounds)	AUTO-	NONE

			PROKARYOTES
			Chlorobiaceae, org. comp.[1]
			    Chromariaceae, org. comp.[1]
			    Rhodospirillaceae[1]
			    Cyanobacteria, H_2O
			    Rhodomicrobium, C_2, C_3
			        compounds
			    Heliobacteriaceae, org. comp.[1]
			    Halobacteria

CHEMO- (chemical compounds)	LITHO-	AUTO-	PROKARYOTES
			    Sulfide oxidizers
			    Methanogens, H_2
			    Hydrogen oxidizers, H_2
			    Methylotrophs, CH_4
			    Ammonia nitrite oxidizers,
			        NH_3, NO^-_2

tion. Speciation, present only in the nonbacterial world, originated when fused bacteria evolved to become the first protoctists.

All eukaryotes (protoctists, fungi, plants, and animals) evolved from complex unions, ultimately of bacteria. The hypothesis of the origins of nucleated

Table 2.1 (continued)

Energy (Light or Chemical Compounds)	Electrons (or Hydrogen Donors)	Carbon Source	Organisms and Their Hydrogen or Electron Donors
		HETERO-	PROKARYOTES "Sulfur bacteria," S Manganese oxidizers, Mn^{++} Iron bacteria, Fe^{++} Sulfide oxidizers, e.g., *Beggiatoa* Sulfate reducers, e.g., *Desulfovibrio*
	ORGANO-	AUTO-	PROKARYOTES Clostridia, etc., grown on CO_2 as sole source of carbon (H_2, $-CH_2$)
		HETERO-	PROKARYOTES (including nitrate, sulfate, oxygen and phosphate[2] as terminal electron acceptors) PROTOCTISTA (most) [3] FUNGI[3] PLANTS (achlorophyllous) [3] ANIMALS[3]

Note: Devised in collaboration with R. Guerrero, Department of Microbiology, University of Barcelona, Spain.

[1]Organic compounds, e.g., acetate, proprionate, pyruvate.

[2]Detection of phosphine: I. Dévai, L. Felföldy, I. Wirner, and S. Plósz. 1988. New species of the phosphorous cycle in the hydrosphere. *Nature* 333:343–45.

[3]Oxygen as terminal electron acceptor.

cells is within a decade of definitive proof of all its postulates. Descriptions of my "serial endosymbiosis theory" (SET) are available (Margulis 1993, 1996). The current status for each organelle is listed in Table 2.2.

The creative force of symbiosis did not end with the evolution of the earliest nucleated cells. Some of the many, often beautiful examples of evolution by symbiosis can be mentioned: *Mixotricha paradoxa*, Pacific coral reefs, New England lichens, New Guinea ant plants, and cows serve as extant examples

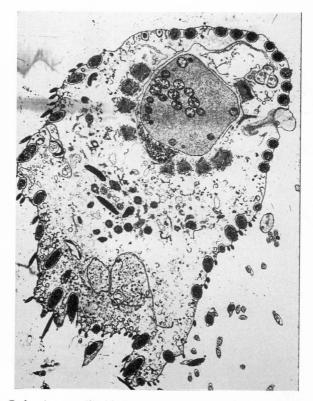

Figure 2.1. *Caduceia versatilis*, hindgut symbiont of *Cryptotermes cavifrons*. This protist harbors at least four distinct bacterial symbionts: two on its surface, one in its cytoplasm, and one in its nucleus. Transmission electron micrograph; reprinted with permission from the *European Journal of Protistology* 35:332 (1999).

of the power of living fusions (Figure 2.2). Members of different species (and sometimes even different kingdoms) under identifiable stresses formed tightly knit communities. These evolved by body fusion and genetic integration into more complex individuals. Individuality arose from community on many occasions (Table 2.3).

SEX AND PARASEX

I will identify and reference here four well-documented but poorly known cases in which microbial and other mergers account for the origins of species.

Table 2.2
Symbiogenetic Origin of Eukaryotic Cell Organelles

Organelle[1]	Source	Status	Reference
Nucleus	Recombinant from *Thermoplasma*-like archaebacteria and *Spirochaeta*-like eubacteria via the organelle karyo-mastigont system	Not proven	Margulis et al. 2000
Undulipodium (cilium, kinetosome)	*Spirochaeta*-like eubacterium via the karyomastigont	Not proven	Margulis 1993 Margulis et al. 2000
Mitochondrion	α-Proteobacteria	Proven	Gray et al. 1999 Margulis 1993
Plastids	*Prochloron*-like cyanobacteria with chlorophylls *a* and *b*	Proven	Margulis 1993

[1] In order of acquisition by hypothesis.

Plant Origins

Some fungi (yeasts, molds, mushrooms) enjoy the power to degrade tough plant material (wood, paper, and other forms of cellulose) and to resist desiccation. A case has been made that all plants evolved from green algae but not directly. Green algae acquired and hid inside their chloroplasts certain wily fungi that conferred cellulase and desiccation resistance on their descendants. Plants evolved from green algae only after the algae incorporated a fungus genome that is present in plant nuclei to this day. The evidence collected by Peter Atsatt (2001) is under investigation.

Speciation in Marine Animals

We describe the origin of larvae (immature forms) of marine animals as formulated by Donald Williamson, Port Erin Marine Station of the University of Liverpool. He strongly infers fertile sexual encounters between adult animals of very different parentage (in some cases they do not even belong to the same phyla). He posits that genetic fusions some 225 million years ago probably

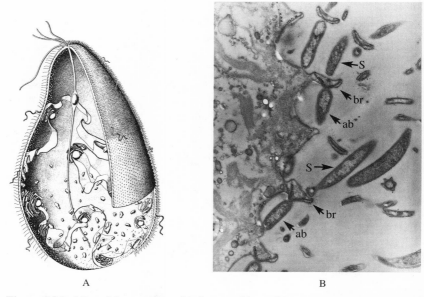

Figure 2.2A. *Mixotricha paradoxa*, hindgut symbiont of *Mastotermes darwiniensis*.
A. Gram negative bacterial epibionts and endobionts, treponeme small spirochetes,
Canaleparolina large spirochete epibiont (Wier et al. 2000), wood particles, endoplas-
mic reticulum, nucleus, axostyle, and four undulipodia can all be seen. Drawing of
the whole cell by Christie Lyons.
Figure 2.2B. Section through *Mixotricha paradoxa* reveals the attached treponema
spirochetes (S), the modified protist surface, and its anchor bacteria (ab). The bract
(br) attaches the epibionts. Figure 2.2A is based on micrographs such as this. Electron
micrograph by A. V. Grimstone.

occurred between echinoderms (such as starfish) and urochordates (such as sea
squirts). Now that molecular biological techniques exist to test it, his theory of
the origin of highly complex developmental pathways from larva to adult de-
serves far more attention than it has received (Williamson 1992, 2001).

Symbiogenesis

Morphogenesis does not occur only because of gene transcription, translation,
and regulation. Impressive morphological discontinuities are induced by symbiogenesis. Productive bizarre parasexual relations with fertile outcomes
occurred several times between land organisms of different species in the evo-

Table 2.3
Examples of Individuality from Community

Kingdom (of Larger Organisms)	Individuals	Community Components	Intra- or Inter-Specific	Communities Classified as Individuals
Protoctist	Amoebae	"Slugs"	Intra-	Cellular slime molds, *Dictyostelium*
Animal	Green algae + *Convoluta* flatworms, diatoms	*Prasinomonas,* heterotrophic *Convoluta,* diatoms	Inter-	*Convoluta roscoffensis Coniventes paradoxia*
Fungus	Yeast + Lactic acid bacteria	More than 20 different bacteria and fungi	Inter-	Kefir granules "Mohammed pellets"
	Trebouxia, or cyanobacteria + ascomycetes	Actinbacteria green algae—cyanobacteria fungi	Inter-	Any lichen, i.e., *Cladonia cistatella, Xanthoria parietina*

lutionary history of life. The fusion of a soil fungus and a nitrogen-fixing filamentous bacterium to form the "bladder plant" *Geosiphon pyriforme* still happens once every three or four weeks during the growing season of this "no-common-name" monster. Green sloths, weevils (beetles), and whales among the animals, and *Gunnera*, cycads, and legumes among the plants, by their very existence illustrate the origins of both large and small species by inherited genome acquisition.

Primate Origins

Chromosome changes in large mammalian clines correlate with species. In descriptions of mammals, especially lemurs of Madagascar and the monkeys and apes of Africa and Asia, the biologists Neil Todd (2000) and Robin Kolnicki (2000) assembled the single most comprehensive and plausible account of the relation between chromosome change and species origin. This research, an entire body of literature on the animals, their current and fossil distribution patterns, their chromosomes and mating habits, compiled by these researchers and others, supports the "karyotypic fission theory" of animal speciation. The mechanism, linked to the ultimate bacterial origins of cell functions such as motility, mechanosensitivity, and reproduction by binary fission, is known as kinetochore reproduction theory (Todd 2001). This example, taken from the primate evolution literature, explains the pattern of speciation of mammals and their chromosome numbers that tend from small to large. Former bacteria that became parts of cells more than 2 billion years ago never went away; they still persist. Again, life remembers its past crises.

CHALLENGES FOR A NEW CENTURY: THE LANGUAGE OF EVOLUTION

The language of evolution sometimes seems to obfuscate more effectively than it illuminates. I anticipate the language to be more chemical, observational, and closer to the live beings. Will biologists drop financial terms (cost, benefit, spend, disadvantage) and simple mathematical symbols for relationships (+ for symbiosis, – for parasitism) and replace them with more adequate descriptions? The adequacy of the current neo-Darwinist theory to explain the origins of new, heritable features of life and of new species has never been shown. The neo-Darwinists who rely on accumulation of random mutations in DNA are not so much wrong as oversimplified and incomplete. Will we see the reinstatement, in the light of new knowledge, of the Darwinian, not the neo-Darwinian, concept of evolution as the organizing principle for the understanding of life? The language of evolutionary change is neither mathematics nor computer-generated morphology. Rather, natural history, ecology, genetics, and metabolism of the large organisms must be supplemented with accurate knowledge of microbes. Will microbial physiology and ecology be seen as essential to an understanding of the evolutionary process? The behavior of microbes within their own populations and in their interactions with others determined life's evolutionary course. The living subvisible world ultimately underlies the behavior, development, ecology, and evolution of the much larger world of which we are a part and with which we coevolved.

ACKNOWLEDGMENTS

I would like to thank Michel Dolan, Ugo d'Ambrosio, William Frucht, and Ricardo Guerrero for helpful discussion, NASA Space Sciences, the University of Massachusetts Graduate School, and the Lounsbery Foundation for funding support, and Jennifer Benson, Judith Herrick, and Donna Reppard for assistance in preparation of the manuscript.

LITERATURE CITED

Atsatt, P. R. In prep. The mycosome hypothesis: Fungi propagate within plastids of senescent plant tissue.

Balows, A., et al., eds. 1992. *The prokaryotes*. New York: Springer-Verlag.

De Kruif, P. 1953. *Microbe hunters*. New York: Harcourt, Brace.

Dobzhansky, T. 1973. Nothing in biology makes sense except in the light of evolution. *Am. Biol. Teacher* 35:125–29.

Gray, M. W., G. Burger, and B. F. Lang. 1999. Mitochondrial evolution. *Science* 283:1476–81.

Kolnicki, R. 2000. Kinetochore reproduction in animal evolution: Cell biological explanation of karyotypic fission theory. *Proc. Natl. Acad. Sci. USA* 97:9493–97.

Margulis, L. 1993. *Symbiosis in cell evolution*. 2d ed. New York: W. H. Freeman.

———. 1996. Archaeal-eubacterial mergers in the origin of Eukarya: Phylogenetic classification of life. *Proc. Natl. Acad. Sci. USA* 93:1071–76.

Margulis, L., M. F. Dolan, and R. Guerrero. 2000. The chimeric eukaryote: Origin of the nucleus from the karyomastigont in amitochondriate protists. *Proc. Natl. Acad. Sci. USA* 97:6954–59.

Margulis, L., and D. Sagan. In press. *Origins of species: Inheritance of acquired genomes*. New York: Basic Books.

NHK-TV. 1990. *Wonders of the rainforest*. Osaka: NHK.

Sonea, S., and L. G. Mathieu. 2000. *Prokaryotology: A coherent view*. Montreal: Les Presses de l'Université de Montreal.

Todd, N. B. 2001. Kinetochore reproduction underlies karyotypic fission theory: Possible legacy of symbiogenesis in mammalian chromosome evolution. *Symbiosis* 29:319–27.

Wier, A. M., J. Ashen, and L. Margulis. 2000. *Canaleparolina darwiniensis,* gen. nov., sp. nov., and other pillotinaceous spirochetes from insects. *Intl. Microbiol.* 3:213–23.

Williamson, D. I. 1992. *Larvae and evolution*. New York: Chapman & Hall.

———. 2001. Larval transfer and the origins of larvae. *Zool. J. Linn. Soc.* 131:111–22.

3

Bodies and Body Plans, and How They Came to Be

MARVALEE H. WAKE

Morphology and development are in many ways both the oldest and the newest facets of biology. The morphology of organisms is what we perceive—it is the phenotype, with all its variation and different expressions. Morphology is shape, and size, and color, and ornamentation; it is body parts—stems and leaves, limbs and jaws, sperm, neurons. Morphology is the structural basis for organismal function (i.e., physiology), including respiration, metabolism, and reproduction, and in many organisms—animal, plant, and "microorganism"—for locomotion and other forms of behavior. It is both the proximal contact and the facilitator of an organism's interaction with its environment. Morphology is the structural composition of organisms at many levels of the hierarchy of biological organization—molecules to cells to tissues to organs to individuals. Morphology is also ontogeny, that is, the physical representation of the changes in structural and functional arrangements that organisms undergo during their lifetimes—from fertilization and cleavage to an adult state, often the reproductive phase, and then the changes of adulthood through senescence to death. The study of morphology can be descriptive, comparative, mechanistic, experimental, and evolutionary.

Development is the way morphology comes to be, in both ontogenetic time and evolutionary time. Development is the process by which morphology is achieved; it changes over ontogenetic time in individuals and over evolutionary time in species and higher taxa; it is shaped genetically and ecologically; and it provides both variation and constraint. Analysis of development, too, includes many levels of the biological hierarchy, ranging from the signaling molecules and proteins that initiate development to the interaction of development with physical factors such as temperature and light. The scientific study of development today is largely experimental and mechanistic, but it is founded in description and can incorporate comparison and even evolutionary frameworks.

One of the great revelations of later twentieth-century biology is of the common genetic basis for much of the morphological organization of organisms—microbial, plant, or animal—and at the same time the basis for the interactions that produce organismal diversity through development and over ontogenetic and evolutionary time scales. Therefore bodies and body plans will loosely be the theme of this chapter—morphology is the pattern, the expression, the phenotype; development is the process by which it is achieved, through genetic and environmental mediation. Bodies and body plans, how they come to be, developmentally and evolutionarily, how they change, and especially what understanding these and related questions will contribute to resolution of some of the scientific challenges of the next century are the questions I will address.

The recently celebrated wedding of development and evolution has attracted much attention because of the insights that developmental biology provides on micro- and macroevolution. (However, given my biases, I believe that more explicit attention should be paid to the way morphology has contributed to that wedding, and the way the wedding is modifying morphology.) Development and evolution are currently enjoying a splendid honeymoon, with major scientific advances daily and the approbation of scientists and the public, after a long courtship that was at times divisive and difficult. But I expect that in the near future the marriage will become polygamous, recognizing morphology and incorporating ecology, behavior, and so on in order to understand and appreciate the bases of complexity and change of organisms and of ecosystems. An integrative approach to the study of morphology and development will contribute to a more synthetic paradigm for the study of biology as a science. That study promises new contributions to society, including the abilities to repair developmental abnormalities, manufacture new body parts from biomemetic materials, make robots that can go places living organisms cannot, understand biodiversity and its maintenance and value, and commit to the social and economic importance of understanding the basis of life.

A HISTORICAL PERSPECTIVE

The last century has witnessed great change in the science of morphology and development. The century opened with developmental biology being largely descriptive embryology, practiced by some of the great morphologists of all time (e.g., Weismann, von Baer, and Haeckel)—those who raised the curtains on the plays of experimental biology and evolutionary biology. But shortly after the turn of the century, morphology and development diverged intellectually and theoretically, as morphology became a major player in systematic biology and evolutionary theory, and development became an experimental field. Both, in various ways and at various times, incorporated new discoveries, in theory and techniques, from genetics, biochemistry, and the emerging fields of cell and molecular biology. By the end of the twentieth century morphology had also become an experimental field, while retaining its descriptive and analytical framework; it now incorporates techniques from other scientific arenas, including genetics, ecology, physics, and engineering. We have a new synthesis of development and morphology in many ways, as questions about the evolution of developmental patterns and processes, the ways morphologies arise and change, and the ways morphologies interact with one another and the environment are attracting the attention of many biologists. The new polygamy is generating new directions for research and discussion.

I will give just a few examples of contributions in morphology and development that illustrate the progression of our understanding from the turn of the nineteenth to the twentieth century to the present, in the context of our understanding of bodies and body plans, and that open a window on the future of the science and its contributions. Then I will conjecture about the future—where I hope development and morphology will progress, how I hope they will be major parts of a new, integrative approach to questions of reductionism and complexity, and how they will contribute to new science and to society as a whole, with an ethic that emphasizes development, morphology, variation, structure-function relationships, and their conservation and potential in the broadest sense. My examples are of animals, mostly vertebrates, because I know them best, but the general principles I will develop obtain for other organisms as well, and my perspective is that of an organismal, evolutionary, and integrative biologist.

Developmental biology became an experimental field as it diverged from descriptive morphology and embryology at the beginning of the twentieth century. The stage had been set by the emergence of questions that could not be answered by inference from dissections and histological examination. The sequence of development (not the stages) of many animals, invertebrate and vertebrate, and plants had been described in detail. Significant similarities in early development, followed by great morphological divergence, had been mapped. Weismann's theory of the continuity of the germ plasm was being explored, and von Baer, Haeckel, and others had promulgated various versions of the biogenetic theory (Nordenskiöld 1928; Hall 1999)—summarized in the notion that ontogeny recapitulates phylogeny, or that stages of development within and across lineages are more similar in early development than in later. These ideas reinforced the concept of the Bauplan, or body plan, and illustrated that much of variation is modification of an overall body plan.

Ernst Mayr (1982) has provided a thought-provoking analysis of the recent history of morphology and development, emphasizing the century that followed the publication of Darwin's *Origin of Species* in 1859 but pertinent to the most recent forty years as well. He commented that at the beginning of the twentieth century zoologists continued the search for homology of all components of anatomy, with an emphasis on phylogeny and the reconstruction of common ancestors. This search produced extensive descriptive treatises, often with analyses of transitions in morphologies, such as fins to limbs and reptile jaw components to mammalian middle ear bones. But Mayr noted that post-Darwinian morphologists ignored a major tenet of Darwinian biology—the explanation of adaptation. They were dealing with archetypes, or common body plans, and their explanations.

Mayr attributed a change in paradigm to the development of evolutionary morphology in the 1950s, when morphologists stopped looking backward to the common ancestor but used the common ancestor as the starting point to ask what evolutionary processes were responsible for the divergence of the descendants. The questions became how new morphologies arise, how selection pressures are involved, what are the characteristics of populations in which changes occur. As Mayr put it, "Evolution . . . [is] the totality of the processes that are involved in evolutionary change." He saw evolutionary morphology building bridges to ecology and behavior in order to assess those processes, but he stated that "the solution to perhaps the greatest problem of morphology requires a bridge to genetics, a bridge which at this time cannot yet be built." He was referring to the origin and the meaning of the great anatomical types, already known to Buffon under the name "unity of plan." And he cited the mammalian Bauplan, which includes types as different as whales, bats, moles, horses, and humans "without any essential change of the mammalian plan."

In several of his more recent books Mayr has developed the idea of the domains of the genotype, especially the somatic domain and what happens to it during evolution. He has asked a series of questions (Mayr 1988): "What happens to the genotype during speciation? What happens in the genotype during . . . evolutionary innovations of the phenotype? What structures of the genotype are responsible for long-time stasis, including the preservation in ontogeny of ancestral developmental stages (such as gill arches in tetrapods) and the stability of the Bauplan (common body plan) of the major types of organisms?" He asserted in his book *One Long Argument* (1991) that "development involves highly complex interactions between different domains of the genotype and dif-

ferent somatic programs," and that study of those interactions would be a holistic approach of great potential to understanding the bases of evolution, but he
indicated that such work had not yet begun.

B. K. Hall in *Evolutionary Developmental Biology* (1999) asked similar questions about the developmental processes that initiate and maintain Baupläne
(body plans). Hall defines body plans as common, basic organizational plans
that reflect community of descent, and then modification of structure within the
body plan (concepts going back to Darwin and before). For most workers the
Bauplan is common to higher taxa, but the idea that it is a nested set of body
plans incorporating and characterizing the evolution of lineages is now receiving attention. As Hall (1999) elucidated, any Bauplan beyond the level of the
species represents a nest; for example, snakes have a Bauplan different from
those of lizards and turtles but still share a reptilian Bauplan; reptiles and mammals have individual body plans but share the vertebrate Bauplan. This concept
permits analysis of the origins of similar structures at any taxonomic level, the
search for mechanisms of homology, and the evolution of the origin of characters. The characters of Baupläne can include physiological and ecological as well
as morphological features, so Baupläne are "functional types" (Kaufman 1995).

At the same time investigation of the nature of Baupläne is coupled with phylogenetic analysis, so taxonomic levels and the nature of monophyletic and
polyphyletic groups are essential to comparative analysis. Hall posed six major
questions about body plans: whether Baupläne constrain development and/or
evolution, how Baupläne arose in evolution, how rapidly they were assembled,
why there are so few Baupläne, why no new Baupläne have arisen in the last
half billion years, and whether the existence of body plans necessitates
macroevolutionary processes. He noted that this is the opportunity to ask how
similar structures arise at any taxonomic level. He anticipated a search for the
mechanisms of homology, the unraveling of the roles of key innovations and
constraints in the origins of body plans, and determination of whether characters, features of morphologies, arise sequentially or in a coordinated manner.
These questions are fundamental to much current study of morphology and development in a broad sense and allow the integration of many fields of biological investigation. Hall's *Evolutionary Developmental Biology* presented an eru-
dite history of experimental developmental biology and the recent incorporation
of interest in the evolution of development and development in evolution.

Another notable contribution to the field is Wallace Arthur's *Origin of Animal Body Plans* (1997). Arthur introduced his examination of the origins of
body plans and the evolution of animal development by noting that neo-
Darwinian theory is incomplete because it lacks the component of individual de-

velopment or ontogeny, which is obvious when one reads attempts to explain the origins of the approximately thirty-five animal body plans, and especially the developmental patterns that generate them. Arthur examined development and the morphologies it generates in terms of paleontology, selection theory, and especially developmental genetics. Developmental genetics to him includes the coevolution of interacting genes, gene expression patterns in time and space, and comparisons of patterns (and perhaps processes) across lineages of animals. (Arthur's table of the rules for naming and abbreviating genes and proteins for those organismal biologists who may not have assimilated the acronyms is a welcome contribution.) His discussion of body plans and phylogeny is synthetic; it argues for "bringing back morphology" in that context.

Both Hall and Arthur review developmental mechanisms thoroughly but from different perspectives. However, the arena of developmental genetics is exploding, and new information accumulates daily. As already noted, eight years ago Ernst Mayr proclaimed his ideas about the interactions of genotypic domains and somatic programs "only words." Since then a substantial amount of work has accumulated that has built the missing bridge to genetics. Researchers have begun to work through those "highly complex interactions between different domains of the genotype and different somatic programs" and to elucidate answers to some of the questions that Mayr, Hall, and those before (and after) them have posed, as shown in the following examples.

DEVELOPMENT AND MORPHOLOGY: EXAMPLES OF INTEGRATIVE APPROACHES

I will present four examples to elucidate the progress in understanding what development and morphology are, as both pattern and process, and to illustrate their contributions to the science of the next century. I have chosen these particular examples because they illustrate novel approaches, are far-reaching in implications, and are potentially or actually integrative approaches to the study of morphology and development, and in some cases ecology and evolution. Remember, my perspective is that of an organismal biologist but one who believes in a transhierarchical approach to biological complexity.

Example 1: The Organizer

One of the major accomplishments in developmental biology was Hilde Mangold and Otto Spemann's identification of the dorsal lip of the blastopore as the "organizer" region that somehow controls or facilitates development and differentiation in frogs and salamanders. With a series of experimental implantations of dorsal lip tissue into various regions of the gastrula at slightly different stages of development and differentiation, they demonstrated that the tissue caused the formation of particular structures, especially a second body axis and all its derivatives (e.g., notochord, neural tube, somites), no matter where it was placed (Spemann and Mangold 1924). The dorsal lip tissue as the organizer of vertebrate development quickly became dogma, and the experiments were repeated in different animals, so that a general principle was assumed. However, for many years following the Spemann-Mangold reports there was no information about what the organizer really was and how it implemented and influenced development. In fact, Ernst Mayr noted as recently as 1991, paraphrasing his colleague June Goodfield (1968), that one must avoid the acceptance of unanalyzed terms, such as *organizer,* which only inhibit analysis.

In the decade since that cautionary comment, understanding of the organizer has dramatically increased. Richard Harland identified the gene *noggin* about ten years ago; its product is expressed in the cells of the dorsal lip of the blastopore. Since then he and others have identified a set of genes that are so expressed and have begun working out how they influence organization, induction, and differentiation (see Harland and Gerhart 1997; Gerhart and Kirschner 1997). These genes determine various aspects of head and body development, including the differentiation of neural structures (Table 3.1), as well as those of mesodermal derivatives. For example, Daniel Bachiller and colleagues (2000) showed that *noggin* and *chordin* are responsible for forebrain development in mice. These genes are present in several animals and presumed to be in all vertebrates and probably invertebrates as well; it is becoming clear that their interactions vary among species. It was initially thought that the products directly induce, but recent research shows that the gene products are not directors or inducers of development; rather they are antagonists, or derepressors, so they "permit" development. Such inducers act by forming complexes with signaling molecules to block the binding sites with the receptors of the signals, so that the signals are inhibited (Brunet et al.1998; Mariani and Harland 1998; McMahon et al.1998; Zimmerman, Jesus-Escobar, and Harland 1996).

However, as experiments and data proliferate it is becoming apparent that the timing of turn-on of the genes, the nature of the receptors of the gene products, and the kinds of interactions among the commonly maintained genes vary considerably among taxa of animals, and, therefore, the developmental and morphological variation long observed is being supplied a mechanistic explanation. Although that explanation is not fully complete, developmental biologists are pursuing reductionistic, fine-grained exploration of these mechanisms. At the

Table 3.1
Genetic Products and Proteins and Their Induction Effects in the Organizer

Product/Protein	Induction Effect
Chordin	Neuralizing, dorsalizing
Follistatin	Neuralizing, dorsalizing
Nodal	Neuralizing, dorsalizing
Noggin	Neuralizing, dorsalizing
Sonic hedgehog	Neuralizing

same time developmental biologists are acquiring a new sense of their potential contributions to questions of the evolution of form and function, and the evolution of organisms, which in conjunction with the work of morphologists, evolutionists, systematists, and ecologists should begin to elucidate the nature and effect of variation and change in organisms—that is, the nature of natural selection (e.g., Northcutt 1993, 1995, 1996).

Example 2: Body Plans and Vertebrate Ancestors

The search for the ancestors of vertebrates is a long-standing example of efforts to understand bodies and body plans. In the early twentieth century there were several competing hypotheses about what invertebrate taxon the vertebrate ancestor might have been: Geoffroy St.-Hilaire had advocated insects, Kovalevsky and Garstang tunicate tadpoles, and Dohrn annelid worms (as the ancestors of both arthropods and vertebrates) (summarized in Hall 1999). The annelid and arthropod theories were based on the nature of segmentation, either embryological or adult. Most workers ruled out insects and annelids as vertebrate ancestors simply because they were invertebrates and therefore "separate" (Hall 1999, 84–85). However, Geoffroy St.-Hilaire had had a significant insight in 1822—that insects are upside-down versions of vertebrates, with ventral nerve cords, dorsal hearts, and so on. The idea of the dorsoventral reversal became part and parcel of the annelid ancestor theory as well, despite the difficulty of explaining the development of the notochord, building a new mouth and anus, and so on in vertebrates. The idea was laid to rest for a while as evidence accumulated for the relationship of ascidians to vertebrates, and both to hemichordates, and of these taxa to echinoderms as the nearest "invertebrate" relatives, all based on the key features of deuterostome development—the patterns of the mouth and anus, the coelom, and so on.

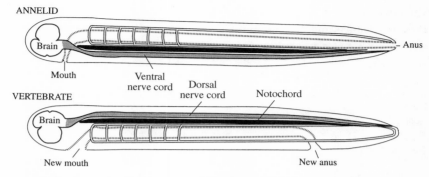

Figure 3.1. The annelid-chordate dorsoventral axis body plans. Is the chordate plan that of an annelid turned upside down? Genetic analysis of axis patterning is shedding new light on an old question (modified after Romer 1960; Hall 1999).

The notion of the upside-down annelid persisted, though, at least as a historical perspective, through the 1960s, as illustrated in the great comparative anatomy textbook by A. S. Romer (1960) and by Hall (1999). There has recently been an interesting resurrection of that idea (see Figure 3.1), based on the homology of the genes that regulate dorsoventral patterning in *Drosophila* and in *Xenopus* (see Arthur 1997). In *Drosophila, decapentaplegic (dpp)* is expressed in a broad dorsal band, and *short gastrulation (sog)* is in a broad ventrolateral band. Similar to *Hox* genes, these genes are involved in the cascade of gene-switching events that determine the spatial arrangement of morphological structures as they develop (François et al. 1994). Further, some of the genes downstream of those are involved in specification of dorsal versus ventral structures, such as *tinman*, downstream of *dpp*, in dorsal heart formation (Azpiazu and Frasch 1993).

Scott Holley and his colleagues (1995) examined the accumulating genetic evidence for patterning in *Drosophila* and *Xenopus*. They suggested that a reversal of the dorsal-ventral axis occurred after the divergence of the common ancestor of insects and vertebrates, based on the patterns of genes with suggestive sequence homology and expression patterns in *Drosophila* and *Xenopus*. Homologues of *dpp* and *sog* in *Drosophila* are the genes *chordin* and *bmp-4* (bone morphogenetic protein) in *Xenopus; bmp-4* is expressed ventrally and *chordin* dorsally, so the pattern is indeed opposite that of *Drosophila*. Holley et al. (1995) confirmed the functional equivalence of the two systems by a series of interspecific injections of the mRNAs for the respective genes. *Drosophila dpp* and *sog* mRNAs were injected into *Xenopus* embryos, *chordin* mRNA from *Xenopus* into *Drosophila*. The *Drosophila* gene products mimicked the effects of the *Xenopus* homologues when injected into the tadpoles (i.e., *dpp* induced ventral rather than dorsal structures, and *sog* dorsalized instead of ventralizing). *Chordin*, when tagged to the N-terminal region of *sog* mRNA, ventralized in *Drosophila*.

Therefore the developmental mechanics of the host embryo are induced by gene products either of their own genes or of homologues from distantly related species. This supports the evolutionary interpretation of Geoffroy St.-Hilaire's theory that axis formation reversed after the divergence of either insects or vertebrates from their common ancestor. Not all workers consider the case for homology and axis inversion conclusive, but many are persuaded. These results do, though, open new questions. For instance, What are the mechanisms by which taxa maintain their evolved pattern responses to genes rather than reverting to the presumed ancestral condition when homologues are introduced?

Lennert Olsson and B. K. Hall (1999) stated that the first great period of understanding of body organization was the late nineteenth century; the second is now. The drivers of that new understanding include molecular phylogenies of the Metazoa that challenge the view of a Cambrian explosion of body plans and indicate that their origins are much earlier in evolutionary time; new ideas about the origin and distribution of patterns of early development; and demonstrations of the sharing of highly conserved homeobox genes, now known to deal not only with specific patterns of segmentation but with much of body plan organization, and of other genes and of signaling molecules that participate in the development of the body plans of many animals (and their homologues in plants).

I add two further components: (1) investigation of new paleontological data, which is changing the way we think about the timing of origins of major taxa and their body plans, and the relationships of major groups of organisms; and (2) biogeography, with emphasis on the processes by which geology and geography change, and thereby affect the evolution and distribution of organisms. Doug Erwin (1999) pointed out that paleontologists have documented the origins of metazoan body plans beginning about 610 million years ago; major environmental changes seem to have triggered the burst of new forms; developmental innovations can be bracketed in time using molecular evidence. He has concluded that as particular developmental patterns are established, they limit subsequent evolutionary trajectories. Still, developmental mechanisms are highly conserved over evolutionary time and taxonomic distances, but taxonomic variation is great.

Ernst Mayr, with his usual perspicacity, anticipated the template for current research on such questions with his idea of the study of the interaction of the domains of the genotype and of whole somatic programs. In *One Long Argu-*

ment (1991) he emphasized that stages of development are targets of selection and that embryonic stages serve as "somatic programs" in development, tending to become highly conserved in evolution. (And, though he did not say so, such conservation might contribute to the limited numbers of body plans that have developed.) Mayr went on to say that conserved stages aid in reconstruction of phylogeny, but evolutionary interpretations are constrained by the extent to which causations of development have been elucidated by embryologists. He explained that "somatic programs" may underlie most instances of recapitulation; if a structure of an ancestor is retained, it may have been preserved by natural selection because it serves as the somatic program for subsequent ontogenetic stages. This retention imposes constraints on evolution, thereby providing resistance to evolutionary change. He asserted that study of domains of the genotype and of whole somatic programs is a holistic approach that demands attention. Mayr foresaw that there would be progress through further analysis, because "development involves highly complex interactions between different domains of the genotype and different somatic programs." Since 1991 there has been extensive progress in understanding the complexity of the interaction of the domains of the genotype, but his prediction continues to generate new and exciting questions. In particular, more comparative assessment of nonmodel organisms, in a phylogenetic context, is beginning to occur, and those approaches should shed light on questions of development, morphology, the environment, and evolution.

Example 3: Morphology and Biomechanics

The analysis of biomechanics, using morphology as the basis for examining how structures function and organisms work to produce behaviors, such as feeding and locomotion in animals and water transport in plants, has been practiced for a long time, but recently the coupling of new equipment and new approaches has revolutionized the field. For most of the twentieth century the ways organisms function were studied usually for single species in an idiographic approach and usually at a particular level of biological organization—mostly whole organisms but often at the bone-muscle interaction level. Models and general principles of several aspects of structure-function relationships have been developed, mostly during the last fifty, or even ten, years. Two examples summarize progress in the understanding of biomechanics—in these cases of animal locomotion—during this past century.

The gifted Eadweard Muybridge, working in the mid- to late nineteenth and early twentieth centuries, pioneered analysis of animal locomotion (Muybridge [1887] 1957). It was his insight that a series of photographs made in rapid suc-

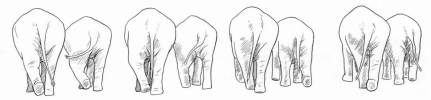

Figure 3.2. Footfall patterns of elephants as an example of the analysis of the biomechanics of locomotion (after Muybridge [1887] 1957).

cession at regulated time intervals would provide the data that would resolve conflicting opinions about animal locomotion, including long-debated questions such as whether a horse ever has all four feet off of the ground, how birds run and fly, how animals shift from walks to trots to gallops—that is, gait analysis. Muybridge used wet collodion plates but was able to get well-resolved images at high frequency and at distance; he developed an automatic exposure system that used motor clocks for making and breaking electrical circuits and the zoopraxiscope, an instrument that demonstrated movements that had been analytically photographed. His apparatus presaged all the image analysis instruments available today, and the same basic principles of photography and analysis guide current work.

Muybridge employed a level of technological innovation paralleled to some degree today in his use of the relatively new field of photography, coupled with a structure that facilitated quantitative analysis, but he also explored the questions of interest with a kind of scholarship rarely pursued today. For example, in his consideration of the gaits of the horse, he compared the depictions in Greek and Byzantine pottery, paintings, and statuary with Roman and later efforts. He made the point that there were many stylistic renditions, most with considerable accuracy regarding the limbs of the horse, but they usually depicted the rider as immovable, unresponsive to the horse's gait—truly inaccurate, as his ([1887] 1957) photographic sequences of horse and rider show. His photographs of animals of diverse sorts (Figure 3.2; elephants), and of humans in different movements and both healthy and infirm children and adults, inform analysis of locomotion to this day. For nearly one hundred years there was little progress in kinematic analysis of locomotion, save that provided by major advances in photographic technique. The movie camera, for example, coupled with the development of strobes and image analysis programs, improved technique but introduced few new principles.

However, during the last twenty or so years analysis of the kinematics and biomechanics of animal locomotion has made use of new technology and ideas.

Michael Dickinson and colleagues (2000) have summarized the complexity of locomotion that emerges from the synergistic interactions among the nervous, muscular, and skeletal systems as well as the physical environment. They note that studies of different locomotor behaviors in a diversity of species are resulting in the elucidation of a few general principles common to most forms of animal locomotion. Integrative approaches that include the reductionistic through the holistic show not only how each component of a locomotor system works but also how they function as a collective whole.

Biomechanics and its practitioners are making major contributions to the emerging field of robotics, and the integration of biological and engineering principles is resulting in new instruments that meet significant social needs. For example, my colleague Robert Full and his associates have looked at the locomotion of cockroaches, centipedes, crabs, salamanders, and other animals (e.g., Martinez, Full, and Koehl 1998; Kubow and Full 1999). They have found that many animals use an alternating tripod gait and have analyzed the mechanics of the system. They have also studied gait changes, direction changes, and the differences in intermittent and sustained locomotion. Understanding of how animals locomote is allowing the development of "walking" robots to investigate uneven substrates, including the bottoms of lakes and oceans. The tripod gait that characterizes most animal locomotion and the "revelation" that animals do not move in straight lines at constant speeds but must adjust to compensate for both external and internal factors are principles that are revolutionizing robotics. Further, the adjustment can be simply a physical property of the appendages of a crab or a robot—neural feedback is not required. Big and small robots are being developed that can explore oceans and go into terrestrial areas where humans cannot (or should not) venture, and miniaturized robots are being developed that can potentially be employed in blood vessels—but making them able to move is the key.

Similarly, Dickinson's work with flight in flies has given us whole new ways of looking at morphology (see Dickinson 1999). For example, it was long thought that a key feature of the evolution of flies, members of the order Diptera, was evolutionary reduction of their pair of hind wings and that the remaining rudiments, the halteres, were vestigial structures. However, Dickinson and his colleagues have shown that the halteres are equilibrium organs that detect angular rotation of the body during flight. They have mapped the haltere-mediated reflexes and determined that haltere afferent nerves provide direct input to a steering motor neuron (Dickinson 1999; Fayyazuddin and Dickinson 1996). Dickinson and his colleagues have developed "robofly" to test models of neural input to control of flight.

Functional morphology and biomechanics are informing engineering, and engineering and the physical sciences are informing morphology. The instruments now available to functional morphologists are far beyond those Muybridge had: much better cameras, computer-aided devices and analysis, treadmills, running tracks, flumes, wind tunnels, and others. Many functional morphologists are becoming highly integrative as they look at the feedback from the skeleton to the nervous system, and muscle fiber dynamics at one level of understanding locomotion, and the mechanical properties provided by the environment at another. As I will discuss, these advances in understanding and application of biological principles can make major contributions to instruments that may work for the good of humanity and of nature—but also possibly increase the rate of nature's destruction.

Example 4: Development, Morphology, and Ecology

A final example deals with the progress in this century from descriptive morphology and development to the integration of studies of development and morphology with ecological and behavioral information in order to understand aspects of the evolution of taxa, including their extinction. Through much of the twentieth century morphologists, embryologists, natural historians, ecologists, and systematists were engaged in the description of the adult morphology of many species, the embryology and development of a few species, and the ecology of even fewer species, mostly in single-species approaches. Only relatively recently has focus on the comparative method employed in a phylogenetic context given morphologists and systematists, and other biologists, new insight into the nature of characters and the relationships among taxa. (Again this is something Ernst Mayr predicted and recommended, as he has kept abreast of the technical and theoretical advances that have made the work possible.) Of course, community ecologists were examining multispecies systems, but usually not with an eye to the evolution of the systems and the relationships of the taxa in them, or to broad-based comparisons. This, too, has changed recently. We gained a splendid base for more complex analysis in all the information gathered about the structure, function, and development of the individuals that compose populations and species. I offer an example of research during the last ten years that I hope might serve as a model for the research of the next century as it explores the complexity of morphology and development in environmental and evolutionary contexts.

Developmental biology, coupled with ecology and life history studies, can tell us about changes in biodiversity. In 1990 two scientists reported finding multilegged frogs and salamanders in and around a small set of ponds in California and suggested a possible mechanism for the development of extra limbs

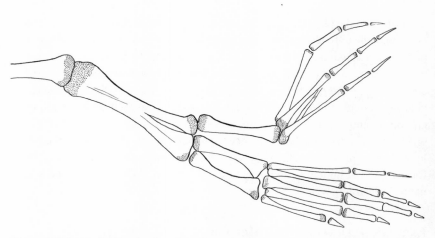

Figure 3.3. Limb duplication in a frog, likely a consequence of interactions of parasites, frog developmental physiology, and stress to the frogs caused by environmental changes that outstrip the compensatory ability of processes of evolution.

(Sessions and Ruth 1990). The ponds had been monitored from 1974 until 1986 because the salamander species, *Ambystoma macrodactylum croceum,* was an endangered taxon; the frogs, *Hyla regilla,* were highly abundant. There were no individuals with extra limbs during the initial twelve years of monitoring. Then in 1987, and massively in 1988, numbers of individuals of both the frog and the salamander were found with extra limbs (Figure 3.3). Among the salamanders, 39 percent of late larvae, 38.5 percent of juveniles, and 4.6 percent of adults had extra limbs; of the frogs, 72 percent of adults had hind limb anomalies and fully 50 percent had extra full limbs, ranging from one extra on one side to nine on the left side and three on the right. There were few front limb anomalies.

Chemical analysis of the pond water revealed nothing unusual. The ecologist S. B. Ruth, who had discovered the multilegged animals, enlisted the aid of a developmental biologist, S. K. Sessions. Sessions quickly recognized that all the specimens of both species were infected with metacercarial cysts of a trematode flatworm parasite, and that there were concentrations of cysts in the tissues at the bases of the hind limbs but also elsewhere. The cysts were in close association with the extra limbs in both the frogs and the salamanders. But the question remained whether the cysts caused the extra limbs or whether they were just infections of already anomalous or sick animals.

Based on knowledge of limb development, Sessions developed a straightforward experiment on lab populations of the frog *Xenopus laevis* and the salamander *Ambystoma mexicanum,* the axolotl, both "model organisms" in developmental biology. He implanted resin beads the size of the metacercarial cysts in the developing limb buds of larvae of the two species; extra limb structures occurred in 20 percent of the experimental animals. Sessions and Ruth (1990) concluded that mechanical disruption of tissue interactions in the limb bud resulted in bifurcation of primordia and developmental duplication of limb structures. This is consistent with what is known of the pattern of extension and bifurcation as limb cartilages aggregate and develop, and the fact that tissue damage during development causes duplication events in a number of animals, including vertebrates and insects. The case is an example of a morphological and developmental perturbation caused by an environmental factor—parasitic flatworms.

Recently populations of frogs with multiple limbs have been found worldwide. There are alternative hypotheses for multiple limbs, such as stimulation or mimicry of the developmental inducer retinoic acid in the animals (Gardiner and Hoppe 1999), but the parasite hypothesis remains persuasive (Johnson et al. 1999; Sessions, Franssen, and Horner 1999). (Parasites are not always present when multiple limbs develop.) None of the hypotheses explains why the amphibians suddenly became susceptible to the parasites—were they more vulnerable because of other environmental stresses? Did the flatworm population suddenly increase? This is an instance of developmental biology shedding light on a morphological and ecological phenomenon that may be a significant warning to all of us about the condition of our environment.

WHAT CHALLENGES IN THE NEXT CENTURY MIGHT THE STUDY OF MORPHOLOGY AND DEVELOPMENT HELP RESOLVE?

There are some general challenges to biology in the next century that are so obvious they almost defy articulation: (1) answering the unanswered questions, and posing new ones; (2) integrating the study of development and morphology, and other subfields, into a framework that generates clear principles of biology; and (3) making the contribution of biology to society clear and relevant. How does the study of development and morphology, and the application of the data and analyses that emerge from the study, relate to these general challenges? As I have illustrated with the preceding examples, morphology and development are alive and well as fields of biology participating in a new synthesis and integration so that issues of biological complexity can be understood. The study of morphology and development, at any and all levels but especially in an integrative manner that looks at multiple parts, multiple effects, and multiple associations of structure and function, has unlimited potential to contribute to understanding of ourselves and the world around us.

Developmental biology is opening up whole new vistas of science. Its integration of genetics with biochemistry is providing new understanding of the structural elements and the morphologies and functions that are part of development and ontogenetic change. Developmental biology is opening the black box of how morphologies unfold and are maintained in individuals, and to some degree in species, by providing a mechanistic approach to the iteration of morphology. Developmental biologists are giving evolutionary biologists new ways to consider the relationships of taxa to one another by revealing the genetic and developmental patterns and processes that are common to many taxa.

Major unanswered questions, though, remain, including those with which I started this discussion: Why are there so few body plans? What provides variation among taxa and individuals, given that much of the regulation of development depends on a common substrate? How does the genome respond to the environment through development? As Hall (1999) noted, there is not a one-to-one correspondence of genotype to phenotype. Therefore developmental biologists must integrate environmental factors in order to understand what might alter the developmental pattern to provide variation. We still do not know how new morphologies, including "key innovations," arise, and we do not know why distantly related organisms respond with similar morphologies to certain environmental parameters (homoplasy, or convergent and parallel evolution; see Sanderson and Huffard 1996). An understanding of how current body forms are produced and how they vary should yield new knowledge of how to produce new body forms—that is, speciation—and also an understanding of why old body forms no longer "fit" their environments—that is, extinction. Lineages do not just die off all at once, but developmental biology might give us answers about why the genotype and the variation expressed are not sufficient to allow organisms to adapt to change in the environment.

Developmental biology also has the potential to make major contributions to the understanding of biodiversity generation and maintenance, and of the current decline in biodiversity. An example of such research is the work of Andrew Blaustein and his colleagues (Blaustein et al. 1994, 1998, 1999; Fite et al. 1998; Hayes et al. 1996; Kiesecker and Blaustein 1995, 1999). They have demonstrated that excess UVB dramatically affects hatching success of the embryos of several species of frogs and toads in Oregon; they propose that UVB is destroying the embryos' DNA, and that their levels of the repair enzyme photolyase cannot cope with the rate of destruction. The increase in UVB levels, presumably caused by human effects that have altered the ozone layer, has happened so rapidly that the repair system cannot evolve in response, and the populations of the animals are in serious decline. Blaustein and his colleagues are also investigating the synergistic effects of pathogens, particularly species of fungi, in these vulnerable animals.

Blaustein's integrative approach is a model, because he continues to expand his combination of ecology, biochemistry, and development to try to understand the phenomenon of population declines. New appreciation for the interaction of mechanisms of development and environmental cues and signals should give insight into interactions and maintenance, as well as the effects of disruption of the environment on developmental patterns and potentially on evolution and extinction.

For morphologists the new insights into developmental constraint and mechanism should allow new paths for analysis of major questions such as the way adaptation, in both evolutionary and physiological senses, occurs; whether behavioral and ecological changes drive morphological change or morphological variation drives behavioral and ecological changes (and the mechanisms by which either might occur). Recent developments in systematic and evolutionary biology, and in engineering and biophysics, allow new ways of examining questions of structure-function relationships and how organisms really work, ranging from the subcellular mechanisms through integrated examinations of neural control and environmental mediations.

Morphologists have more to say now about the nature of characters for use in systematic analysis, and the issues of functional units, character independence, and so on take on new meaning. But I do urge them to view their research more reciprocally than most now do—what functional morphology tells us about structure and thereby characters should be communicated to systematists and evolutionary biologists so their database is increased and ever more robust phylogenetic hypotheses can be generated. Similarly, the understanding of phylogeny developed by systematists and evolutionists should be better communicated to other biologists. Use of the comparative method is enjoying a renaissance (Harvey and Pagel 1991; Brooks and McLennan 1991), and new methods of phylogenetic analysis—for the production of robust hypotheses on the relationships of taxa to determine the direction of evolutionary change—are facilitating that renaissance.

New awareness of the patterns of morphology and the processes of development that underlie the behavior and ecology of organisms is arising. That recognition occurs, for example, as ecologists learn that the species composition, rather than just the activity, of "functional groups" or guilds matters (see Tilman 1999; Symstad et al. 1998) and behaviorists see the interplay of structure at all levels, such as the biochemistry of color of a feather, the structure of its parts, and the neuromusculoskeletal basis for waving feathers to produce a behavior that elicits a response from a viewer or potential mate.

Integrative analysis of development and morphology, when meshed with other subfields of biology and with information from the physical and social sciences, can lead to understanding of the origin and evolution of biological complexity at several levels. We need to know how complexity arises, functions, and is maintained at subcellular through organismal through ecological and behavioral levels; we need to know how complexity, including patterns and processes of development and morphologies, or organisms, can change. Developmental mechanisms provide some of that understanding. Analysis of developmental and morphological mechanisms will provide answers to other kinds of questions of the evolution of complexity. For example, it is becoming apparent that small genetic changes direct major morphological changes of various sorts, ranging from sexual dimorphism to insect castes. Understanding of how such change is directed is likely to facilitate understanding of the evolution of complex behaviors and even social systems.

But I have so far alluded only to some of the general research questions that remain unresolved. Developmental biology and morphology are also making significant contributions to society, ranging from medical to ecological applications, and their potential for further contribution is ever expanding. Understanding of developmental mechanisms, especially their genetic basis, is facilitating the initiation of new treatments and even corrections of heritable diseases and deformities. Craniofacial development is a case in point—understanding the biological basis of the development (and evolution) of the head is providing new understanding of the bases for anomalies of the teeth, the jaws, the musculature, and so on, and therefore means of modulating those anomalies either developmentally or surgically. Research is also determining that some developmental anomalies can be prevented by a healthy diet, and therefore might not require genetic manipulation or surgical intervention—the reduced incidence of spina bifida, or the failure of completion of the neural arches of vertebrae in humans and other vertebrates, by including high levels of biotin in the maternal diet is an example. The biological basis of such structural modification needs more extensive exploration.

Bioengineering is dealing with materials for repair—the skeleton of corals is now being used to assist bone repair; functional morphologists and biomechanicians are providing the principles and the empiricisms that lead to development of better prosthetic devices. I have already mentioned advances in robotics—devices that can go where humans cannot, whether on ocean substrates or in blood vessels, are being developed, and only the imagination limits the possibilities for their employment. Other uses of morphology—devices that can

read the retinal image of the eye or the print of the hand to be used as an identification marker, for example—are being developed. The potential contributions of developmental biology and morphology to the aid of society are unlimited—if a focus on the quality of life of the donor species, as well as the
recipients, is maintained.

We are doing less well in conveying the potential applications of our advances in developmental biology and morphology to other socially important
questions. For example, we constantly observe ontogenies—children growing
up, tadpoles becoming frogs, caterpillars becoming butterflies, seedlings becoming flowers or herbs or trees. But we concentrate too often on single species,
and think in terms of stages, not continua; we forget about aging too—it is the
intergenerational interaction of all these components that generates our environments. The species composition of the "stages" of ecological succession is in
effect ontogeny at several levels—that of the individual organisms as well as that
of the ecosystem over time. We are literally seeing the trees but not the forest.

To carry that allusion further, I think of ecology as the study of the interaction of morphologies of different species with one another and with the physical environment. Even the functional groups so popular in ecological analysis
today could profitably be characterized by the morphological features that provide the functional contribution to the interactions of the species involved. Such
a perspective might further elucidate the composition of functional groups in
like habitats in different regions and shed light on the concept. Morphological
features such as tree height, leaf form, or the morphologies of top carnivores
are major features of the organisms' function in their habitats—morphology is
what they are and where they are, and development is how they got that form.

Getting back to a more holistic view, I suggest that a more integrative, transdisciplinary, multihierarchical approach to hypothesis formation, testing, and
analysis of major questions and problems will provide a much better understanding of biology and its contribution to society. We need to develop our research enterprise in terms of significant, synthetic themes that will affect many
parameters of science and society.

We must recognize that many elements of the progress in development and
morphology have been aided immeasurably by advances in technology, as illustrated by the examples I have cited. Computer-aided data generation and
analysis, modeling and simulations, search engines, and so on have made science work faster and more innovatively. Instrumentation of all sorts is making
new analyses possible. Technology clearly will continue to provide new devices
and new applications, but all of them will depend on the ideas of the inventors
who devise them and the scientists who use them to make progress, and both
scientists and designers must be responsible for wise application of technology.

The twenty-first century is sure to be an exciting time for developmental biologists and morphologists. The advances in technology will facilitate basic research and its applications to important questions in biology and its social dimensions. Innovative ideas abound. There appears to be vast potential for development of good science. More and more people with both skills and ideas are being trained. However, some of them cannot get jobs, and the distribution of scientists, geographically, nationally, and in kinds of research, is skewed. What do we need to continue and accelerate new technology and new ideas, and to incorporate a view that science operates for the good of society? We need education systems that have a well-grounded philosophy about the teaching of science and then the capacity to do it well. We need teachers confident of their knowledge of the science they are teaching, including efficient means for them to update their information. We need teachers who go beyond the facts, who can be integrative and transdisciplinary, so that their students will be able to understand science broadly.

Such systems and teachers will provide a more scientifically literate and open-minded public. Our constituency would then be well acquainted with not just the facts of science but the way it is done. It would likely press for the kind of science that takes a broad view, and it would understand the contribution of science to the good of society. A public educated with a synthetic perspective will also be aware that the good of society is not simply the good of humankind but is a product of understanding the implications of all the interactions that affect all life on this planet. Advancing the understanding of morphology and of development, preferably together, will be a useful part, but merely a part, of good, integrative, progressive science. We need people with the vision of an Ernst Mayr in classrooms at all levels of instruction, at the bench, and in the field as scientists—the hard part is achieving the uniqueness of Ernst Mayr, but the transmission of his vision, and visions like it, is one way to begin.

CHALLENGES FOR A NEW CENTURY

I have elucidated many new questions and problems, and some unsolved old ones, in this discussion. As techniques and instrumentation are developed, and as scientists recognize the contribution toward resolution that a synthetic and integrative approach will provide to complex problems, the science of the new century has exceptional promise for dealing with significant problems. Examples of the kinds of complexity that can be understood through new ways of examining problems, including those of the development, morphology, evolution, and ecological interactions of the organisms that inhabit the Earth, include the following:

1. Use of the techniques and approaches of developmental biology and developmental biologists, geneticists, and morphologists to analyze the evolution of new body forms and new species, and contribution of phylogenetic analysis to understanding patterns and processes of development
2. Understanding of the interaction of structure and function, as well as their evolution, through hierarchical and integrative approaches that include multidimensional analyses and the resources of nonbiological (e.g., chemical, physical, engineering) interpretations
3. Examination of the way environmental factors affect development and the morphologies that arise, in terms of "stages" of both life histories and whole ontogenies, especially the consequences to fitness of individuals and populations
4. Application of growing understanding of the interactions of genes, cells, organisms, species, and environments to questions of viability, extinction, conservation, and management (when necessary and appropriate) of life on the planet
5. Education of our intelligent public and policy makers about the nature and contribution of an understanding of biology, including development and morphology, in a computer-literate but increasingly specialized and urban world

We have the resources of technique, innovative colleagues, and a renewed interest in biological science to consider questions of significance that are both intellectual and pragmatic. In fact, an approach to studying questions in biology that meshes the philosophical and the pragmatic has great promise for advances in the new century.

ACKNOWLEDGMENTS

Many students, colleagues, and fellow members of boards and committees have influenced my thinking about development and morphology, the contributions of those fields to biology, and the nature and significance of integrative biology. I appreciate recent discussions with several colleagues, and their loans of reprints and illustrative materials, for the formulation of this contribution. Comments from the reviewers and editors improved the manuscript. I especially thank Adam Summers for preparing several of the slides used in the oral presentation and Karen Klitz of the Museum of Vertebrate Zoology, University of California at Berkeley, for executing the figures. Finally, I am pleased to acknowledge the support of the National Science Foundation for my research in morphology, development, and evolutionary biology.

LITERATURE CITED

Arthur, W. 1997. *The origin of animal body plans: A study in evolutionary developmental biology.* Cambridge: Cambridge Univ. Press.

Azpiazu, N., and M. Frasch. 1993. *Tinman* and *bagpipe*: Two homeobox genes that determine cell fates in the dorsal mesoderm of *Drosophila. Genes & Dev.* 7:1325–40.

Bachiller, D., J. Klingensmith, C. Kemp, J. A. Belo, R. M. Anderson, S. R. May, J. A. McMahon, A. P. McMahon, R. M. Harland, J. Rossant, and E. M. De Robertis. 2000. The organizer factors Chordin and Noggin are required for mouse forebrain development. *Nature* 403:658–61.

Blaustein, A. R., J. B. Hayes, P. D. Hoffman, D. P. Chivers, J. M. Kiesecker, W. P. Leonard, A. Marco, D. H. Olson, J. K Reaser, and R. H. Anthony. 1999. DNA repair and resistance to UV-B radiation in western spotted frogs. *Ecol. Appl.* 9:1100–1105.

Blaustein, A. R., P. D. Hoffman, D. G. Hokit, J. M. Kiesecker, S. C. Walls, and J. B. Hays. 1994. UV-B repair and resistance to solar UV-B in amphibian eggs: A link to population declines? *Proc. Nat. Acad. Sci.* 91:1791–95.

Blaustein, A. R., J. M. Kiesecker, D. P. Chivers, D. G. Hokit, A. Marco, L. K. Belden, and A. Hatch. 1998. Effects of ultraviolet radiation on amphibians: Field experiments. *Amer. Zool.* 38:799–812.

Brooks, D. R., and D. A. McLennan. 1991. *Phylogeny, ecology, and behavior: A research program in comparative biology.* Chicago: Univ. of Chicago Press.

Brunet, L. J., J. A. McMahon, A. P. McMahon, and R. M. Harland. 1998. Noggin, cartilage morphogenesis, and joint formation in the mammalian skeleton. *Science* 280:1455–57.

Dickinson, M. H. 1999. Haltere-mediated equilibrium reflexes of the fruit fly, *Drosophila melanogaster. Phil. Trans. R. Soc. London B* 354:903–16.

Dickinson, M. H., C. Farley, R. J. Full, M. Koehl, R. Kram, and S. Lehman. 2000. Toward an integrative view of how animals move. *Science* 288:100–106.

Erwin, D. H. 1999. The origin of body plans. *Amer. Zool.* 39:617–29.

Fayyazuddin, A., and M. H. Dickinson. 1996. Haltere afferents provide direct, electrotonic input to a steering motor neuron in the blowfly, *Calliphora. J. Neurosci.* 16:5225–32.

Fite, K. V., A. R. Blaustein, L. Bengston, and J. E. Hewett. 1998. Evidence of retinal light damage in *Rana cascadae*: A declining amphibian species. *Copeia* 1998:906–14.

François, V., M. Solloway, M. W. O'Neill, J. Emery, and E. Bier. 1994. Dorsal-ventral patterning of the *Drosophila* embryo depends on a putative negative growth factor encoded by the *short gastrulation* gene. *Genes & Dev.* 8:2602–16.

Gardiner, D. M., and D. M. Hoppe. 1999. Environmentally induced limb malformations in mink frogs *(Rana septentrionalis)*. *J. Exptl. Zool.* 284:207–16.

Gerhart, J., and M. Kirschner. 1997. *Cells, embryos, and evolution*. Malden, Mass.: Blackwell Science.

Goodfield, J. 1968. Theories and hypotheses in biology. *Boston Studies Phil. Sci.* 5:421–49.

Hall, B. K. 1999. *Evolutionary developmental biology*. 2d ed. Dordrecht: Kluwer Acad. Publ.

Harland, R., and J. Gerhart. 1997. Formation and function of Spemann's organizer. *Ann. Rev. Cell Dev. Biol.* 13:611–67.

Harvey, P. H., and M. P. Pagel. 1991. *The comparative method in evolutionary biology*. Oxford: Oxford Univ. Press.

Hayes, J. B., A. R. Blaustein, J. M. Kiesecker, P. D. Hoffman, I. Pandelova, D. Coyle, and T. Richardson. 1996. Developmental response of amphibians to solar and artificial UVB sources: A comparative study. *Photochem. Photobiol.* 64:449–56.

Holley, S. A., P. D. Jackson, Y. Sasai, B. Lu, E. M. DeRobertis, F. M. Hoffman, and E. L. Ferguson. 1995. A conserved system for dorsal-ventral patterning in insects and vertebrates involving *sog* and *chordin*. *Nature* 376:249–53.

Johnson, P. T. J., K. B. Lunde, E. G. Ritchie, and A. E. Launer. 1999. The effect of trematode infection on amphibian limb development and survivorship. *Science* 284:802–4.

Kaufman, D. M. 1995. Diversity of new world mammals: Universality of the latitudinal gradients of species and Bauplans. *J. Mammal.* 76:322–34.

Kiesecker, J. M., and A. R. Blaustein. 1995. Synergism between UVB radiation and a pathogen magnifies amphibian embryo mortality in nature. *Proc. Nat. Acad. Sci.* 92:11049–52.

———. 1999. Pathogen reverses competition between larval amphibians. *Ecology* 80:2442–49.

Kubow, T. M., and R. J. Full. 1999. The role of the mechanical system in control: A hypothesis of self-stabilization in hexapedal runners. *Phil. Trans. R. Soc. London B* 354:849–61.

McMahon, J. A., S. Takada, L. B. Zimmerman, C.-M. Fan, R. M. Harland, and A. P. McMahon. 1998. Noggin-mediated antagonism of BMP signalling is required for growth and patterning of the neural tube and somite. *Genes & Dev.* 12:1438–52.

Mariani, F. V., and R. M. Harland. 1998. XBF-2 is a transcriptional repressor that converts ectoderm into neural tissue. *Development* 125:5019–31.

Martinez, M. M., R. J. Full, and M. A. R. Koehl. 1998. Underwater punting by an intertidal crab: A novel gait revealed by the kinematics of pedestrian locomotion in air versus water. *J. Expt. Biol.* 201:2609–23.

Mayr, E. 1982. *The growth of biological thought: Diversity, evolution, and inheritance*. Cambridge, Mass.: Harvard Univ. Press, Belknap Press.

————. 1988. *Toward a new philosophy of biology: Observations of an evolutionist.* Cambridge, Mass.: Harvard Univ. Press, Belknap Press.

————. 1991. *One long argument: Charles Darwin and the genesis of modern evolutionary thought.* Cambridge, Mass.: Harvard Univ. Press.

Muybridge, E. 1957. *Animals in motion.* New York: Dover Publ. Reprint of E. Muybridge. 1887. *Animal locomotion.* Philadelphia: Univ. of Pennsylvania Press.

Nordenskiöld, E. 1928. *The history of biology: A survey.* New York: Tudor Publ.

Northcutt, R. G. 1993. A reassessment of Goodrich's model of cranial nerve phylogeny. *Acta Anat.* 148:71–80.

————. 1995. The forebrain of gnathostomes: In search of a morphotype. *Brain Beh. Evol.* 46:275–319.

————. 1996. The origin of craniates: Neural crest, neurogenic placodes, and homeobox genes. *Israel J. Zool.* 42:273–313.

Olsson, L., and B. K. Hall. 1999. Introduction to the symposium: Developmental and evolutionary perspectives on major transformations in body organization. *Amer. Zool.* 39:612–16.

Romer, A. S. 1960. *The vertebrate body.* Philadelphia: Saunders.

Sanderson, M. H., and L. Huffard, eds. 1996. *Homoplasy: The recurrence of similarity in evolution.* San Diego: Academic Press.

Sessions, S. K., R. A. Franssen, and V. L. Horner. 1999. Morphological clues from multilegged frogs: Are retinoids to blame? *Science* 284:800–802.

Sessions, S. K., and S. B. Ruth. 1990. Explanation for naturally occurring supernumerary limbs in amphibians. *J. Expt. Zool.* 254:38–47.

Spemann, O., and H. Mangold. 1924. Über Induktion von Embryonalanlagen durch Implantation artfremder Organisatoren. *Arch. Mikr. Anat. u. Entw. Mech.* 100:599–638.

Symstad, A. J., D. Tilman, J. Willson, and J. M. H. Knops. 1998. Species loss and ecosystem functioning: Effects of species identity and community composition. *Oikos* 81:389–97.

Tilman, D. 1999. The ecological consequences of changes in biodiversity: A search for general principles. *Ecol.* 80:1455–74.

Zimmerman, L. B, J. M. Jesus-Escobar, and R. M. Harland. 1996. The Spemann organizer signal *noggin* binds and inactivates bone morphogenetic protein 4. *Cell* 86:599–606.

4

Ecosystems

Energetics and Biogeochemistry

GENE E. LIKENS

"Things change" is a predominant paradigm of the last half century. Only brief reflection will call to mind the many major technological changes during the past fifty years (e.g., commercial jet aircraft, color television, E-mail, satellite weather observation). Not only do we rely on these new systems in our everyday lives but increasingly we take them for granted. For example, we become very impatient and irritated when the computer is "down" or slow. Undoubtedly more important than these technological changes is the fact that the human population recently (1999) exceeded 6 billion, increasing by about 3.5 billion, or more than 2.4-fold greater than it was in 1950! The activities of this single species are now dominating ecosystems throughout the globe, altering global biogeochemical cycles of carbon, sulfur, nitrogen, and other elements (e.g., Kovda 1975; Likens 1991, 1992, 1994; Galloway, Levy, and Kasibhatla 1994; Vitousek 1994; Postel, Daily, and Ehrlich 1996; Vitousek et al. 1997a, 1997b), and in other ways markedly changing ecosystem structure and function (Kovda 1975; Stumm 1977; Turner et al. 1990; Likens 1994; Vitousek et al. 1997b).

Many important ecological changes arise from human dominance of the planet, the results of which may be observed in a decade, a year, or even a few months, mere fractions of an evolutionary second. Ecosystems become windows through which patient observers and researchers can reveal these important changes and seek to explain their significance within the larger global patterns and cycles on which life depends.

While some biologists rise to the challenge of unraveling the extraordinary complexity at the molecular level of the genome, ecosystem ecologists seek to understand the functioning of entire ecosystems and landscapes, inhabited by countless numbers of interacting abiotic and biotic entities. Frequently the actions of humans, be they near or far, have a powerful impact on ecosystem functioning and change. The growth of human populations, coupled with modern technology and consumption levels, places humans in the position of dominating ecosystems globally in a way never before seen.

Major new advances and ways of thinking, which clarify this window of ecosystem ecology, including the overarching lenses of ecological energetics and biogeochemistry, have occurred during the past fifty years. Some are indeed new, but I want to stress at the outset that, as is universally the case in science, we all rely today on the foundations provided by those who have gone before in terms of ideas, concepts, principles, and questions. Often what are new are the technological tools and facilities that provide new approaches to and simplification of long-standing, difficult questions.

Excellent examples of "old now new" or continuing questions for ecosystem ecologists are provided by the early writings of Aldo Leopold (1939):

To the ecological mind, balance of nature has merits and also defects. Its merits are that it conceives of a collective total, that it imputes some utility to all species, and that it implies oscillations when balance is disturbed. Its defects are that there is only one point at which balance occurs, and that balance is normally static. [balance of nature—flux of nature]

Land then, is not merely soil; it is a fountain of energy flowing through a circuit of soils, plants, and animals. Food chains are the living channels which conduct energy upward; death and decay return it to the soil. The circuit is not closed; some energy is dissipated in decay, some is added by absorption, some is stored in soils, peats and forests, but it is a sustained circuit, like a slowly augmented revolving fund of life. [food web analysis]

Things change! Currently there is much written prompting a widely held opinion that there have been great improvements in environmental quality in the United States, particularly since the 1970s (e.g., Easterbrook 1995). Indeed, there have been significant improvements (e.g., less obvious air pollution from particulates, less raw sewage in waterways, and less litter along roadways), but there also are major environmental problems and degradation that have emerged since the 1960s and early 1970s (Table 4.1).

My colleagues and I at the Institute of Ecosystem Studies (IES) attempted to evaluate how the environment in the Hudson River ecosystem has changed since the 1970s. We determined that there have been major improvements (less visible trash, less input of raw sewage) but there also have been continuing, new, or worsening problems (e.g., invasion of the alien zebra mussel; Strayer et al.

Table 4.1

Major Emerging Environmental Issues During 1950–2000 with Strong Impacts on Ecosystem Structure, Function, and Change

---

Severe air pollution (e.g., London, 1952)

Widespread application of pesticides (e.g., DDT, Carson 1962)

Toxic chemical events (e.g., methylmercury, Minimata, Japan, 1959; methyl isocyanate, Bhopal, India, 1984; Love Canal, New York, 1978)

Oil spills (e.g., Torrey Canyon, 1967)

Acid rain (1970s to present; e.g., Odèn 1968; Likens et al. 1972)

CFC-induced ozone depletion (e.g., Crutzen 1971; Molina and Rowland 1974;Rowland and Molina 1975)

Land-use changes (e.g., shrinking of Aral Sea from water diversion for irrigation, 1970s to present)

Tropical deforestation (1970s–1980s to present; e.g., Myers 1980; Richards 1990)

Invasion of alien species (e.g., zebra mussels into the Laurentian Great Lakes, 1980s)

Nuclear accidents (e.g., Chernobyl, 1986)

Resource depletion (e.g., collapse of Atlantic cod fishery, 1990s)

Climate change (e.g., severe El Niños, such as in 1997–98)

Note: Modified from Munn, Whyte, and Timmerman 1999.

1996). A clear conclusion was that environmental impacts have changed, but, on balance, it would appear to me that this River's overall environmental status has deteriorated in spite of targeted improvements, increased public awareness, and major efforts to protect and restore habitats. It is very difficult to characterize environmental problems as good or bad, better or worse, but clearly things change.

It is extremely difficult to speculate about what challenges and questions ecosystem ecologists will face in the future as unquestionably new materials and new technologies will be developed to provide new perspectives as well as new environmental problems. But if the previous fifty years are any guide, some of the same difficult conundrums (e.g., effects of air pollution, land-use changes, and global climate change; see Table 4.1) will continue to provide ecosystem ecologists with challenges for decades to come.

Thus, in the next few decades we will need to deal with the complex environmental issues that we know about now. In addition, there will be surprises that will enhance the need to develop new knowledge about affected ecosystems (and landscapes) and to guide solutions. In my opinion there will be many, some serious, environmental problems for ecosystem ecologists to tackle in the upcoming decades associated with technological "advances," such as intensive agriculture (e.g., high-density domestic animals; Batie 1993), indiscriminate or excessive antibiotic use (e.g., Mallin 2000), and transportation associated with urban sprawl. If ecosystem ecologists do not monitor and experiment in order to unravel the complexities of ecosystems, and particularly what humans are doing to them, who will?

In any event, my challenge is to review briefly some of the major research foci and achievements in ecosystem ecology during the past fifty years and to identify or describe some of the major challenges and frontiers in the next. Obviously, this is a very difficult task. I will focus here primarily on terrestrial, freshwater, and coastal marine ecosystems.

THE PAST

The Ecosystem Concept

In 1935 A. G. Tansley introduced the term *ecosystem,* stressing the interactions among living and nonliving components constituting a system or patch of nature (Tansley 1935). However, it was not until the appearance of the second edition of Eugene P. Odum's textbook *Fundamentals of Ecology* in 1959 that the concept and approach really took off in popularity and acceptance. I was a graduate student when this book appeared. The ecosystem concept certainly excited me and in many ways contributed to the development of the Hubbard Brook Ecosystem Study in the White Mountains of New Hampshire. The term and the concept were obscure then; now *ecosystem* has become a household word (see Likens 1998).

There are several reasons for Odum's success: his enthusiasm, his clarity of purpose in showing the value of quantitative, large-scale studies, his application of the concept to both aquatic and terrestrial ecosystems, and his approach for understanding complexity at large scales. The ecosystem concept provides a comprehensive framework for study of the interactions among individuals, populations, and communities and their abiotic environments, and for study of the change in these relationships with time (Likens 1992). In my opinion, the development of the ecosystem concept and approach, and its use in understanding and resolving complex environmental problems, is one of the major advances in biology during the past century.

Moreover, I believe that the *ultimate* challenge for Ecology overall is to integrate ecological information available from all levels, approaches, and scales of inquiry into an understanding that is meaningful and useful to managers and decision makers (Likens 1992). This challenge is particularly clear for ecosystem ecologists. Unfortunately, the field of Ecology is now highly fractionated regarding ideas, concepts, and particularly approaches, and it is difficult to bring

an integrated, comprehensive view to bear on complex environmental or resource management issues. Yet we must! Moreover, a split has now occurred
between more biologically focused ecologists (e.g., population ecologists) and
those who focus on the physical-chemical environment (e.g., biogeochemists).
Short-term success with more reductionist approaches clearly leads to more
fractionation (Likens 1992). Nevertheless, integration and synthesis are key to
developing broad ecological understanding. Now, with the opportunity of addressing even larger-scale issues (e.g., current initiatives of the National Science Foundation on Biocomplexity in the Environment [Colwell 2000] or the
National Ecological Observatory Network), it will be incumbent upon ecologists—particularly ecosystem ecologists—to integrate rather than fractionate
and to cooperate even more widely, bringing social, economic, and cultural dimensions and approaches into this integration. Such comprehensive integration
remains an overriding goal for ecosystem science.

The Ecosystem Approach

I have not attempted a comprehensive listing of the major concepts and research
foci of the past fifty years because this has been done elsewhere (e.g., McIntosh
1985; Hagen 1992; Golley 1993; Pace and Groffman 1998). I have assembled
a few highlights of the last one hundred years to indicate the richness and
breadth of these efforts (Table 4.2). I did attempt to list the ecosystem papers
designated by *Current Contents* as Citation Classics, but because this information was not readily available, I compiled instead a list of ecosystem papers that
were included in various "Readings" selected by others as major contributions
to the field during the past sixty or so years (Table 4.3).

The early period in the development of the ecosystem approach can be characterized as a "black box" approach to the study of large systems. This approach
was exciting and powerful as applied to large, complicated ecosystems, such as
a forest or a lake. Inputs, outputs, and "budgets" of energy and chemicals were
determined for units of the landscape—microcosm units, entire lakes, springs,
fields, watersheds—but because of the difficulty of the task, it usually was done
for only one parameter or element at a time. The black box approach helped to
pinpoint important questions or critical mechanisms for an ecosystem and at the
same time provided perspective for why the mechanism was important (e.g.,
impact on adjacent ecosystems or global cycles). Boundaries are, therefore, central to this approach and key for determining quantitative mass balances (see
Likens and Bormann 1985; Likens 1992 for details).

Early work on energetics and trophic structure (Elton [1927] 1939; Juday
1940; Lindeman 1942), ecological pyramids of numbers, biomass, and energy

Table 4.2
Development of Some Major Ecosystem Concepts and Research Foci During the Twentieth Century

The Ecosystem Concept[1]
The Ecosystem Approach
  Application /Utilization of the Concept
    Whole-ecosystem manipulations (experiments)[2]
      Hubbard Brook Experimental Forest
      Coweeta Hydrologic Laboratory
      Experimental Lakes Area
    Long-term (sustained) ecological research[3]
      Legacies, long response times
      Monitoring
    System/Ecosystem modeling[4]
  Research Topics Contributing to Ecosystem Ecology
    Ecosystems at landscape to global scales
      Linking of fluxes among air-land-water systems[5]
    Disturbance and recovery[6]
    Top down–bottom up controls on ecosystem function[7]
    Species and ecosystems[8]
      Alien species invasions, conservation, hindcasting unmeasured variables
        from paleoecological models
    Humans as components of ecosystems[9]
Overarching Themes
  Energetics (energy flux and budgets for ecosystems)
    Energy budgets[10]
    Trophic structure, food chains, food webs[11]
    Ecological stoichiometry[12]
  Biogeochemistry (flux and cycling of chemicals for ecosystems)[13]
    Bio in biogeochemistry[14]
      Radioactive and stable isotopes[15]
    Small watershed approach[16]
      Mass balance
    Nutrient loading models (eutrophication)[17]

---

[1]Tansley 1935; Evans 1956; E. P. Odum 1959.
[2]Hasler, Brynildson, and Helm 1951; Likens et al. 1970; Schindler 1973; Likens 1985; Swank and Crossley 1988; Carpenter et al. 1995.
[3]Likens 1989.
[4]H. T. Odum 1960; E. P. Odum 1964; Van Dyne 1966; Watt 1966; Botkin, Janak, and Wallis 1972; see Lauenroth et al. 1998.
[5]Likens and Bormann 1974b; see Hasler 1975.
[6]Barrett 1968; E. P. Odum 1969; Bormann and Likens 1979; Pickett and White 1985.
[7]Hrbácek et al. 1961; Shapiro 1979; Carpenter and Kitchell 1993.

(continued)

Table 4.2 (continued)

---

[8]See Jones and Lawton 1995.
[9]Stearns and Montag 1974; Turner et al. 1990; see McDonnell and Pickett 1993.
[10]Juday 1940; H. T. Odum 1956; Gosz et al. 1978.
[11]Elton [1927] 1939; Lindeman 1942; H. T. Odum 1957; Slobodkin 1962.
[12]Redfield 1958; Reiners 1986; Sterner et al. 1997; Elser et al. 1996; Elser and Urabe 1999.
[13]Vernadsky 1944, 1945; Hutchinson 1950, 1957; see Burke, Lauenroth, and Wessman 1998.
[14]Redfield 1958.
[15]See Schultz and Klement 1963.
[16]Bormann and Likens 1967.
[17]Vollenweider 1968; Schindler 1977; Vitousek and Howarth 1991; see Smith 1998.

(Odum 1959), and efficiency of energy flow between trophic levels (Odum 1957) for ecosystems laid the foundation for ecosystem research during the next several decades. H. T. Odum's well-known energy analysis of the Silver Springs, Florida, ecosystem (1957) was an early example. Several of these early efforts showed the critical role of microorganisms within ecosystems and the fact that their activities far exceeded their ambient biomass relative to this relation for other trophic levels. Likewise, early work clearly showed the relation of biology to biogeochemistry (Redfield 1958) and provided a foundation for managing eutrophication in freshwater ecosystems (Vollenweider 1968).

The black box approach provided answers to certain important questions. Moreover, large-scale studies including experimental manipulation of entire ecosystems, such as those we initiated at the Hubbard Brook Experimental Forest in 1965, often produced results that could not have been predicted, at least not easily, from small-scale studies. For example, our experimental manipulations of entire watershed-ecosystems showed that clear-cutting of the forest caused large losses of nitrate and calcium in drainage water (Likens et al. 1970). These results clearly pointed to process-level questions inside the box, such as the role of nitrification in these ecosystems (Likens, Bormann, and Johnson 1969; Bormann and Likens 1979). In general, the ecosystem approach has been very efficient and effective in defining and focusing key process-level research within the black box (e.g., nitrification in disturbed forest ecosystems).

David Schindler and colleagues (1985) showed that food-web disruption, which occurred early in the experimental acidification of an entire lake ecosystem in Ontario, could not have been predicted from small-scale (laboratory or microcosm) studies because of the complex and redundant interactions among the various components of the natural ecosystem, which were responding to multiple stresses and diverse changes.

Table 4.3

Papers Selected in "Readings" for Ecosystems, Ecological Energetics, and Biogeochemistry before 2000

1935–50	Tansley, A. G. 1935. The use and abuse of vegetational concepts and terms. Ecology 16:284–307. (II) Pearson, O. P. 1948. Metabolism and bioenergetics. Scientific Monthly 56:131–34. (V)	Juday, C. 1940. The annual energy budget of an inland lake. Ecology 21:438–50. (I) Hutchinson, G. E., and V. T. Bowen. 1950. Limnological studies in Connecticut, 9. A quantitative radio-chemical study of the phosphorus cycle in Linsley Pond. Ecology 31:194–203. (VIII)	Leopold, A. 1941. Lakes in relation to terrestrial life patterns. In A symposium on hydrobiology. Madison: Univ. of Wisconsin Press. Pp. 17–22. (I)	Lindeman, R. L. 1942. The trophic-dynamic aspect of ecology. Ecology 23:399–418. (I, II, V)	Clarke, G. L. 1946. Dynamics of production in a marine area. Ecol. Monogr. 16:321–35. (I)
1951–60	Hayes, F. R., et al. 1952. On the kinetics of phosphorus exchange in lakes. J. Ecol. 40:202–16. (VIII) Redfield, A. C. 1958. The biological control of chemical factors in the environment. Amer. Sci. 46:205–21. (I)	Evans, F. C. 1956. Ecosystem as the basic unit in ecology. Science 123:1127–28. (I) Pomeroy, L. R., and F. M. Bush. 1959. Regeneration of phosphate by marine animals. Intern. Oceanog. Congr. Preprints, 893–94. (VIII)	Odum, H. T. 1956. Primary production in flowing waters. Limnol. Oceanog. 1:102–17. (V) Ryther, J. H. 1959. Potential productivity of the sea. Science 130:602–8. (V)	Rigler, F. H. 1956. A tracer study of the phosphorus cycle in lake water. Ecology 37:550–62. (VIII) Odum, H. T. 1960. Ecological potential and analogue circuits for the ecosystem. Amer. Sci. 48:1–8. (VII)	Odum, H. T. 1957. Trophic structure and productivity of Silver Springs, Florida. Ecol. Monogr. 27:55–112. (I) Pomeroy, L. R. 1960. Residence time of dissolved phosphate in natural waters. Science 131:1731–32. (VIII)

1961–70				
Engelmann, M. D. 1961. The role of soil arthropods in the energetics of an old field community. Ecol. Monog. 31:221–38. (V)	Odum, E. P. 1962. Relationships between structure and function in ecosystems. Japanese J. Ecol. 12:108–18. (I)	Teal, J. M. 1962. Energy flow in the salt marsh ecosystem of Georgia. Ecology 43:614–24. (II)	Margalef, R. 1963. On certain unifying principles in ecology. Amer. Nat. 97:357–74. (I, IV)	Olson, J. S. 1963. Energy storage and the balance of producers and decomposers in ecological systems. Ecology 44:322–31. (IV)
Olson, J. S. 1963. Analog computer models for movement of nuclides through ecosystems Radioecology, 121–25. (VIII)	Johannes, R. E. 1964. Phosphorus excretion and body size in marine animals: Microzooplankton and nutrient regeneration. Science 150:28–35. (VIII)	Beeton, A. M. 1965. Eutrophication of the St. Lawrence Great Lakes. Limnol. Oceanog. 10:240–54. (VII)	Brooks, J. L., and S. I. Dodson. 1965. Predation, body size, and composition of plankton. Science 150:28–35. (II)	Gates, D. M. 1965. Energy, plants, and ecology. Ecology 46:1–13. (IV)
Baker, H. G. 1966. Reasoning about adaptations in ecosystems. BioScience 16:35–37. (IV)	Cole, L. C. 1966. Man's ecosystem. BioScience 16:243–48. (VII)	Engelmann, M. D. 1966. Energetics, terrestrial field studies, and animal productivity. Adv. Ecol. Res. 3:73–115. (IV)	Paine, R. T. 1966. Food web complexity and species diversity. Amer. Nat. 100:65–75. (II, IV)	Sawyer, C. N. 1966. Basic concepts of eutrophication. J. Water Pollution Control Fdn. 38:737–44. (VII)
Bormann, F. H., and G. E. Likens. 1967. Nutrient cycling. Science 155:424–29. (VII)	Patten, B. C., and M. Witkamp. 1967. Systems analysis of ^{134}cesium kinetics in terrestrial microcosms. Ecology 48:813–24. (VIII)	Riley, G. A. 1967. Mathematical model of nutrient conditions in coastal waters. Bull. Bingham Oceanog. Coll. 19:72–80. (VIII)	Davis, M. B. 1969. Climatic changes in southern Connecticut recorded by pollen deposition at Rogers Lake. Ecology 50:409–22. (II)	Odum, E. P. 1969. The strategy of ecosystem development. Science 164:262–70. (II, VII)
Schultz, A. M. 1969. A study of an ecosystem: The Arctic Tundra. In The ecosystem concept in natural resource management, ed. by G. Van Dyne. New York: Academic Press. Pp. 77–93. (VIII)	Likens, G. E., et al. 1970. Effects of forest cutting and herbicide treatment on nutrient budgets in the Hubbard Brook watershed-ecosystem. Ecol. Monog. 40:23–47. (II, VIII)			

Table 4.3 (continued)

1971–80	Walsh, J. J., and R. C. Dugdale. 1971. A simulation model of the nitrogen flow in the Peruvian upwelling system. Investigacion Pesquera 35:309–30. (VIII)	Caperon, J., and J. Meyer. 1972. Nitrogen-limited growth of marine phytoplankton: 1. Changes in population characteristics with steady-state growth rate. Deep-Sea Res. 19:601–18. (VIII)	Johannes, R. E., et al. 1972. The metabolism of some coral reef communities: A team study of nutrient and energy flux at Eniwetok. BioScience 22:541–43. (VIII)	Johnson, P. L., and W. T. Swank. 1973. Studies of cation budgets in the southern Appalachians on four experimental watersheds with contrasting vegetation. Ecology 54:70–80. (VIII)	Jordan, C. F., and J. R. Kline. 1972. Mineral cycling: Some basic concepts and their application in a tropical rain forest. Ann. Rev. Ecol. System. 3:33–50. (VIII)
	Bormann, F. H. 1976. An inseparable linkage: Conservation of natural ecosystems and the conservation of fossil energy. BioScience 26:754–60. (VI)	Bormann, F. H., and G. E. Likens. 1977. The fresh air–clean water exchange. Nat. Hist. 86:63–71. (VI)	Campbell, R. 1977. The interaction of two great rivers helps sustain the Earth's vital biosphere. Smithsonian, Sept. 1977. (VI)	Mortimer, C. H. 1978. Props and actors on a massive stage. Nat. Hist. 87:51–58. (VI)	
1981–90	Allen, T. F. H., et al. 1984. Interlevel relations in ecological research and management: Some working principles from hierarchy theory. USDA Forest Service General Technical Report RM-110, July 1984. (III)	Romme, W. H., and D. H. Knight. 1982. Landscape diversity: The concept applied to Yellowstone Park. BioScience 32:664–70. (III)	Carpenter, S. R., et al. 1985. Cascading trophic interactions and lake productivity. BioScience 35:634–39. (III)	Pastor, J., et al. 1988. Moose, microbes, and the boreal forest. BioScience 38:770–77. (III)	Wiens, J. A. 1989. Spatial scaling in ecology. Functional Ecol. 3:385–97. (III)
	Vitousek, P. M. 1990. Biological invasions and ecosystem processes: Towards an integration of population				

1991–2000	Daily, G. C., and P. R. Ehrlich. 1992. Population, sustainability, and Earth's carrying capacity. BioScience 42:761–71. (IX)	Holling, C. S. 1992. Cross-scale morphology, geometry, and dynamics of ecosystems. Ecol. Monog. 62:447–502. (IX)	Odum, E. P. 1992. Great ideas in ecology for the 1990s. BioScience 42:542–45. (IX)	Costanza, R., L. Wainger, C. Folke, and K.-G. Maler. 1993. Modeling complex ecological economic systems: Toward an evolutionary, dynamic understanding of people and nature. BioScience 43:545–55. (IX)	Jones, C. G., et al. 1994. Organisms as ecosystem engineers. Oikos 69:373–86. (IX)
	Schneider, E. D., and J. J. Kay. 1994. Life as a manifestation of the second law of thermodynamics. Math and Computer Modeling 19:25–48. (III)	Hunsaker, C. T., and D. A. Levine. 1995. Hierarchical approaches to the study of water quality in rivers. BioScience 45:193–203. (III)	Pickett, S. T. A., and M. L. Cadenasso. 1995. Landscape ecology: Spatial heterogeneity in ecological systems. Science 269:331–34. (III)	Risser, P. G. 1995. Biodiversity and ecosystem function. Conservation Biology 9:742–46. (IX)	Holling, C. S., and G. K. Meffe. 1996. Command and control and the pathology of natural resource management. Conservation Biol. 10:328–37. (III)
	Vitousek, P. M., et al. 1997. Human domination of Earth's ecosystems. Science 277:494–99. (III)	Peterson, G., et al. 1998. Ecological resilience, biodiversity and scale. Ecosystems 1:6–18. (III)			

Note: Roman numerals refer to the "Readings" books that selected these papers. I: Kormondy, E. J. 1965. *Readings in ecology.* Englewood Cliffs, N.J.: Prentice-Hall. II: Real, L. A., and J. H. Brown, eds. 1991. *Foundations of ecology: Classic papers with commentaries.* Chicago: University of Chicago Press. III: Dodson, S. I., et al., eds. 1999. *Readings in ecology.* New York: Oxford University Press. IV: Boughey, A. S., ed. 1969. *Contemporary readings in ecology.* Belmont, Calif.: Dickenson. V: Hazen, W. E., ed. 1964. *Readings in population and community ecology.* Philadelphia: W. B. Saunders. VI: Crane, J., et al., eds. 1979. *Readings in ENVIRONMENT 79/80.* Guilford, Conn.: Dushkin. VII: Boughey, A. S., ed. 1973. *Readings in man, the environment, and human ecology.* New York: Macmillan. VIII: Pomeroy, L. R., ed. 1974. *Cycles of essential elements (Benchmark papers in ecology).* Stroudsburg, Pa.: Dowden, Hutchinson, and Ross. IX: Samson, F. B., and F. L. Knopf, eds. 1996. *Ecosystem management: Selected readings.* New York: Springer-Verlag.

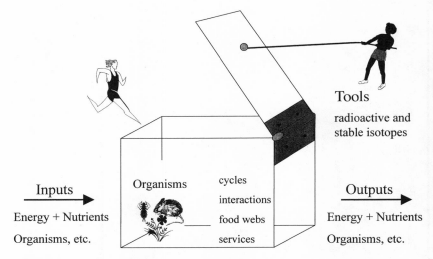

Figure 4.1. Opening the "black box" of ecosystem science.

With time this "black box" has been opened more widely, using large-scale experiments and sophisticated tools such as radioactive and stable isotopes to reveal connections, interactions, multiple stresses, and multiple-element interactions (Figure 4.1). In turn, these internal components and interactions have been opened and explored relative to the ecosystem and landscape matrix. Biological relationships were elaborated through biogeochemical studies, and food webs were identified and unraveled (see Tables 4.2 and 4.3). Surprisingly in hindsight, the role of species needed to be reasserted in ecosystem analyses (Jones and Lawton 1995), and humans had to be put inside the "box." Clearly, individual species, frequently referred to as keystone species (e.g., Power et al. 1996), can have critical roles in the structure and function of ecosystems; a narrow view of considering only "trees" or "green slime" in ecosystem analysis can miss clues important to greater understanding. For example, tabonuco trees (*Dacryodes excelsea* Vahl) help maintain hillslope stability and reduce erosion on steep slopes in tropical rain forests of Puerto Rico because of their unique intraspecific habit of root grafting and root anchorage to subterranean rocks and boulders (Basnet et al. 1992). Three species of snails (*Euchondrus albulus* Mousson, *E. desertorum* Roch, *E. ramonensis* Granot) in the Negev Desert greatly accelerate weathering and nitrogen cycling by eating endolithic lichens, which grow inside the surface layers of rocks (Shachak, Jones, and Brand 1995; Jones and Shachak 1990).

But it is not just species. There are countless additional entities composing ecosystems (including ions, compounds, individuals, and communities), all interacting, processing, aggregating, separating, living, dying, and changing in mind-boggling arrays of complexity. Yet there are measurable, important emergent properties, such as biological productivity, decomposition, and nutrient cycling that provide integrative insights into this complexity, even for very large and structurally diverse ecosystems.

The direct effect of humans on ecosystems (e.g., pollution) had been known and studied for a long time, but understanding the role of humans as components of ecosystems is a more recent effort (Stearns and Montag 1974; McDonnell and Pickett 1993). Humans have both direct and subtle effects on the structure, function, and development of ecosystems, and the effect of human activity can now be observed on ecosystems throughout the globe. So issues like conservation and restoration cannot be addressed realistically in isolation from the human dimension within ecosystems (McDonnell and Pickett 1993; Pickett et al. 1997).

A critical contribution of the past has been the beginnings of integration and synthesis of information for large systems. Individual airsheds interact among themselves as do watersheds, and airsheds interact with watersheds and so forth. Their individual inputs and outputs also are important connections to larger, global cycles (Bormann and Likens 1967; Likens et al. 1977). These larger-scale interactions are less well quantified.

Systems thinking and modeling were early features of the ecosystem approach (e.g., Odum 1964; Van Dyne 1966; see McIntosh 1985). The interconnected flows of organisms, nutrients, energy, and so on as Coupled Ecological Flow Chains represent a new and exciting approach to such analysis of ecosystem function (Shachak and Jones 1995). Examination of the role of organisms as ecosystem engineers (Jones, Lawton, and Shachak 1994) is another important approach for addressing this complexity inside the box and its relation to ecosystem inputs and outputs. Stephen Carpenter (1998) has used the metaphor of a table sustained by the four legs of theory, experiments, comparisons, and long-term studies (Figure 4.2) to summarize the development of ecosystem science.

CHALLENGES: PRESENT AND FUTURE

I have listed in Table 4.4 some major issues and challenges for ecosystem ecology now and in the future. This listing is a simple attempt to identify some of the critical challenges, new tools for addressing complex questions, and major resource management and education issues at the beginning of the twenty-first century for ecosystem ecology. Here I will briefly expand on a few.

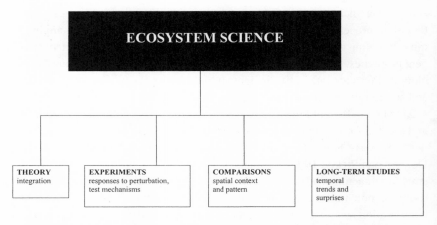

Figure 4.2. Ecosystem science is supported by theory, experiments, comparisons, and long-term studies (modified from Carpenter 1998).

Evaluate Complexity and Linkages Among Air-Land-Freshwater-Marine Ecosystems to Guide Integrated Management Solutions for Landscape to Global Scales for Environmental Problems

There are many examples of complex interactions in ecosystem analysis. I will briefly describe here one related to biogeochemical cycles affecting the anthropogenic alteration of sulfur and nitrogen cycles that produce "acid rain" and one related to the complex interactions among several components of northeastern U.S. forest ecosystems leading to Lyme disease in humans.

Large quantities of sulfur and nitrogen oxides are released to the atmosphere through the combustion of fossil fuels. These elements return to the Earth's surface as strong acids in rain, snow, sleet, and hail, in cloud and fog water, or as acidifying gases and particles. These inputs can acidify aquatic and terrestrial ecosystems. When this problem was first identified in the late 1960s to early 1970s (Odèn 1968; Likens, Bormann, and Johnson 1972; Likens and Bormann 1974a), it was thought to be primarily atmospheric deposition of sulfuric acid that acidified lakes and streams and killed fish (Figure 4.3A). It now is clear that the ramifications of the human disturbance of both the sulfur and nitrogen cycles related to acid rain are much more complex (Figure 4.3B). Specifically, cycles of aluminum, mercury, lead, phosphorus, calcium, magnesium, and carbon may be altered as well, leading to accelerated release of toxic metals, eutrophication of coastal estuaries, and other ecosystem alterations (e.g., Likens 1998).

A superb example of unraveling ecosystem complexity, as well as successful team development and application can be seen in a recent study led by sci-

Table 4.4
Major Challenges Facing Ecosystems Ecology

Major Challenges

Reach a better understanding of complexity

    Combine approaches from molecular to landscape scales

    Combine approaches for evaluating long-term processes and events

    Combine biological-chemical-physical, socioeconomic, and cultural approaches

    Understand and manage urban agglomerations as ecosystems

Evaluate interactions among ecosystems (air-land-freshwater-marine) to guide integrated management

Evaluate interactions and linkages across large spatial scales (landscape to global)

    Accelerating and/or degrading biogeochemical cycles

Maintain and strengthen long-term data sets

Reach a better understanding of ecosystem function and change in developing regions of the world

Manage big projects; train team members and promote team building; achieve *inter*disciplinary approach

Use of New Tools

Stable isotopes, Geographic Information System, remote sensing, the Internet, molecular approaches, modeling, computational power, large and longer-term data sets, submersibles, and so on

Conservation/Restoration/Ecosystem Services

Maintain adequate quantity and quality of water

    Salinization of inland waters

    Toxic algal blooms, pathogens

    Impacts of large dams

    Efficiency of use, protection of quality

Minimize effects of land fragmentation and loss of arable land; manage land-use changes

Manage ecosystem effects of alien species invasions

Establish and maintain ecological reserves

Increase ecological literacy at all levels

---

entists from the Institute of Ecosystem Studies on the complex interconnections among oak trees, deer, mice, ticks, Lyme disease, gypsy moths, and humans. The team consisted of members with different subspecialties in Ecology, each bringing special expertise to bear on this complex ecosystem problem.

Abundant acorns from oak trees (*Quercus* spp.) attract white-tailed deer (*Odocoileus virginianus*) to the forest in the northeastern United States. Mast years (with abundant acorn production) occur every three to four years. The deer carry adult black-legged ticks (*Ixodes scapularis*), which drop off and lay eggs in the forest floor. The abundant acorns also attract mice, such as the white-footed mouse (*Peromyscus leucopus*), which rapidly increase in abundance in

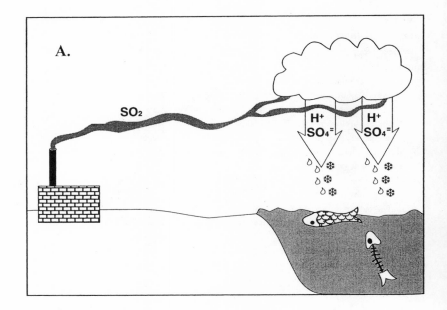

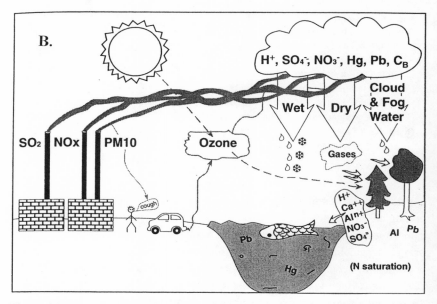

Figure 4.3. Ecosystem response to acid rain. **A.** The "simple" view of the early 1960s to early 1970s. **B.** The increasingly complex view of the 1990s (from Likens 1998).

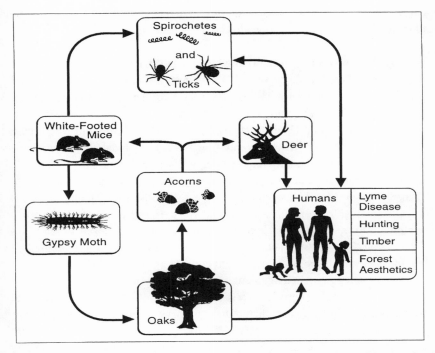

Figure 4.4. Linkages among oak trees, deer, mice, ticks, gypsy moths, and humans in northeastern U.S. forests. Based on Ostfeld et al. 1996; Jones et al. 1998; Ostfeld and Jones 1999. (Reproduced with permission of R. S. Ostfeld and C. G. Jones)

response to the increased food supply. The following summer the tick eggs hatch into larvae, which during the process of obtaining blood meals from mice pick up the spirochete bacterium (*Borrelia burgdorferi*) that causes Lyme disease in humans. The tick larvae transform into nymphs a year or so later and pass on the spirochete when they attach to humans or other mammals for blood meals. In general, risk of Lyme disease in oak forest areas is correlated with mast years of acorns with a lag of two years (Ostfeld, Jones, and Wolff 1996; Ostfeld 1997; Jones et al. 1998; Ostfeld and Jones 1999). Moreover, by predation on the pupae of gypsy moths (*Lymantria dispar*), abundant mouse populations can prevent moth outbreaks that, when they occur, can result in defoliation and tree death in oaks—thus reducing production of acorns and potentially affecting the risk of Lyme disease (Figure 4.4).

Many factors affect our ability to tackle and resolve the complexity of ecosystem structure, function, and change in the future. Here I will discuss briefly three: new tools, team building, and maintenance and use of long-term data, because I believe that these components are fundamental to successful resolution of complex and diverse questions in ecosystem ecology.

New Tools and Integration   Using new technologies, such as remote sensing from satellites, we are becoming able to describe and quantify atmospheric parameters, changes in surficial features, such as dimensions of lakes and glaciers, and biomass and its nutrient content over large areas of the Earth's surface (e.g., Matson and Ustin 1991; Aber et al. 1993; Martin and Aber 1997; Martin et al. 1998; see Wessman and Asner 1998). Such data help clarify the window in posing and addressing new large-scale questions, such as these: Is there a widespread change in the chemistry of vegetation that is related to changes in atmospheric chemistry from large-scale atmospheric pollution? What is the ecological footprint (sensu Rees and Wackernagel 1994) of a large urban agglomeration and associated sprawl within the landscape mosaic of land uses for a region? The massive challenge will be to integrate and apply biological-chemical-physical, socioeconomic, and cultural knowledge to such questions for use in integrated management of large areas (landscapes-watersheds-regions). Commonly, these large areas have a complex, mosaic structure and pulsed human activity and/or disturbance, often occurring at incongruent temporal and spatial scales (Carpenter et al. 2001). This integration will require knowledge and synthesis at the intersections between disciplines, and at large spatial scales, combined with knowledge about the structure, function, and change of ecosystems remote from human activities (e.g., Hedin, Armesto, and Johnson 1995; Galloway et al. 1982).

Unfortunately, there are no precise answers about ecosystem structure, function, and change for large systems because their complexity defies identification of "the answer" with our current tools. Thus, we require high-quality data over larger scales, coupled with new attempts to integrate this diverse information.

Team Building   A critical challenge, if not constraint, as we approach these large-scale problems will be our ability to construct functional and effective multidisciplinary teams (Likens 1992, 1998). Long experience with the Hubbard Brook Ecosystem Study has made it clear to me that forming and nurturing an effective, efficient, and collegial team is a continual challenge. Unfortunately, little formal planning normally is given to this critical task. I predict that as we move to even larger scales and more diverse teams, this challenge will grow commensurately. Previously, I proposed some fundamental characteristics needed in team members for successful team building (Table 4.5) and suggested some specific measures to enhance team building for ecosystem analysis (Likens 1998). The measures include (1) training of team leaders; (2)

Table 4.5

Characteristics of People Needed to Build Successful Teams in Ecosystem Science

Brightness
Ability to trust and be trustworthy (trust)
Abundant common (or good) sense
Creativity and willingness to share with team
Appropriately trained
Collective ability to make up deficiencies
    Shared experiences
Willingness to give team time
Personality
    Listens
    Enjoys working with others
    Is curious and interested
    Is open to new ideas and approaches
Keeping eyes open (serendipity reigns)
Liking for one another

Note: Modified from Likens 1998.

mentoring by experienced team members; (3) face-to-face communication about team and individual expectations; (4) development of effective and efficient time management for individuals and teams; (5) discussions and clear expectations about responsibilities, priorities, openness, and trust; (6) listing of titles of potential publications and authors, including establishing of the order of authors in advance of drafting papers (this is a good practice at the beginning of each year); (7) honing and appreciation of common sense, based on experience and commitment, which is a major ingredient for successful serendipity; and (8) talented administrative help to facilitate team function and accountability.

In addition to high-technology research and large team efforts into the foreseeable future, there will remain a major need for natural-history observations and investigator-initiated research. Combined, these approaches provide a powerful way of tackling ecosystem complexity. Investigator-initiated research routinely provides the core of new ideas and new information, which bring into sharp focus the potential problem resulting from taking funds to support large-scale initiatives without maintaining this core.

Long-Term Data The value of long-term data and sustained research is clear at all levels and for all of the issues identified in Tables 4.1 through 4.4. The value of collecting long-term data was not always appreciated during the past fifty years, particularly in terms of funding support. At the moment that battle seems to have been won, but there are still major problems regarding what should be measured and how, who should do the measurements, and how these data should be stored and accessed among other things. The issue of property rights to data both within teams and outside teams has become a major concern in recent years. These problems need to be resolved, and we need leadership to launch an enlightened and sustained national monitoring effort.

Mindless data collection usually is of limited value. Rarely have ecologists attempted to formulate carefully which questions should drive large-scale and long-term monitoring efforts. It is difficult to identify the critical questions and to guide long-term monitoring in this way, but we could do much better and be much more cost effective than we have been in the past.

There are at least five ways to obtain or approximate long-term information: (1) direct long-term measurements, (2) retrospective studies, (3) ecological modeling, (4) space-for-time substitution studies, and (5) experimentation (Likens 1989). These approaches rarely are used together, but results are much more powerful and integrative from the combination.

Long-term data are especially valuable in providing reference or baseline information for evaluating trends and for ecosystem experiments, as well as for determining whether events are unusual. Knowledge about legacies and/or slow response times in ecosystems is critical for prediction and management. Long-term data help resolve such questions.

Long-term data on ammonium and nitrate in bulk precipitation and stream water at the Hubbard Brook Experimental Forest have revealed not only interesting patterns in historical disturbance (legacies) but important ecological questions (Figure 4.5). Total dissolved inorganic nitrogen (DIN) input in bulk precipitation increased from 1964–65 to about 1972–73 (water-year extends from June 1 to May 31; Likens and Bormann 1995) and then leveled off at about 500 moles per hectare-year. In contrast, DIN output in stream water increased from 1964–65 to 1969–70, remained relatively high for eight years (1969–77), and then decreased markedly to quite low values with three exceptions, 1979–80 to 1980–81, 1989–90, and 1998–99. The higher output values during 1969–77, 1979–80 to 1980–81, and 1989 may have been due to increased soil frost during those periods, which mobilized nitrate from the soil (Likens et al. 1977; Mitchell et al. 1996; Lewis and Likens 2001). The increase in DIN during 1998–99 was correlated with extensive ice-storm damage to forest vegetation during January 1998, increasing nitrate export in drainage waters.

Forest biomass is a major reservoir of nitrogen in this forest ecosystem, so it is puzzling and counterintuitive that streamwater DIN output did not increase after 1982, when accumulation of forest biomass ceased (see Figure 4.5). These

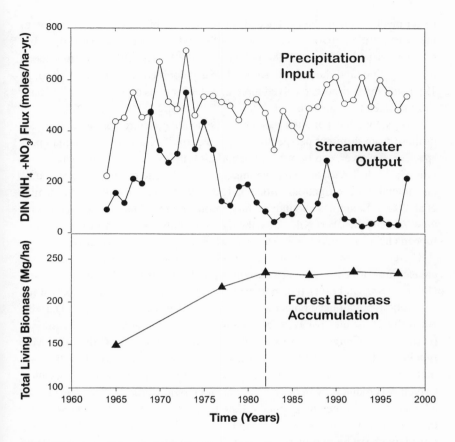

Figure 4.5. Annual fluxes of dissolved inorganic nitrogen (NO_3^- + NH_4^+) in precipitation (○) and stream water (●) and total forest biomass accumulation (▲) for Watershed 6 of the Hubbard Brook Experimental Forest in New Hampshire. (Forest biomass based on Bormann and Likens 1979; Likens et al. 1998; and T. G. Siccama, personal communication.)

complexities and effects of legacies are the subjects of intensive current study. After thirty-seven years of intensive study much ecosystem complexity still has not been unraveled.

Managing the Ecosystem Effects of Alien Species Invasions

Normally, alien species are considered by population and community ecologists; however, invasive alien species can have major impacts on ecosystem energetics and biogeochemistry. The problem of invasive alien species is likely to get much worse in the next decade or so because of its strong connection to the world trade network (see Bright 1999) and the increased ease of global travel. Christopher Bright (1999) points out that 80 percent of the world's goods travel by ship, and the volume shipped nearly doubled from 1970 to 1996. In 1989 only three airports received more than 1×10^6 tons of cargo; by 1996 there were thirteen such airports.

Ecosystem effects of alien species can be positive, neutral, or negative, or all of these from different perspectives. The negative impacts from a human perspective, however, can be very large and difficult to manage ("biological pollution"; Bright 1999), often threatening public health and costing billions of dollars annually for mitigation (Office of Technology Assessment 1993; Pimental et al. 2000). Examples abound: rabbits introduced into Australia, possums into New Zealand, the West Nile virus, the tiger mosquito and gypsy moths, and European cheatgrass into North America, whiteflies into South America.

There are several possibilities for dealing with invasive species, such as increased use of pesticides, stricter control and treatment of ballast water (see Holden 2000) and other transport vectors, genetic modification to control reproduction, but all have potential ecosystem effects. A recent event in Darwin, Australia, illustrates some of the complex issues. An alien dreissenid mussel [*Mytilopsis* (=*Congeria*) sp.] suddenly appeared in Darwin's Cullen Bay Marina in late 1998 (Bax 1999). The waters of Darwin Harbor and all anchorages of the Northern Territory were examined carefully for the presence of this alien (which was found only in three locations of Darwin Harbor). It was decided to add enough chlorine and copper sulfate to the Bay to kill 100 percent of the mussels. This treatment was done and judged effective (Bax 1999), but there must have been numerous known and unknown ecosystem effects. The decision to act swiftly and decisively is unusual in such matters but may reflect Australia's past bad experience with alien species. Nevertheless, obvious questions remain concerning the trade-offs between the ecosystem effects of this invasive species and the ecosystem effects of the treatment, and whether or how often this episode will need to be repeated. What are the short-term and long-term economic and ecologic costs and benefits?

The invasion of zebra mussels (*Dreissena polymorpha*) into North America provides a good case study. These mussels were first seen in the freshwater section of the Hudson River ecosystem in 1991 (Strayer et al. 1996, 1999; Caraco et al. 1997). This alien mussel increased in abundance rapidly and by 1993 was the dominant filter feeder in the ecosystem. On average the huge population of zebra mussels in the Hudson River theoretically filter the entire freshwater portion of the River every 1.2 to 3.6 days (Strayer et al. 1999). Fortunately, basic components of this ecosystem had been studied since 1986 (e.g., Cole et al.

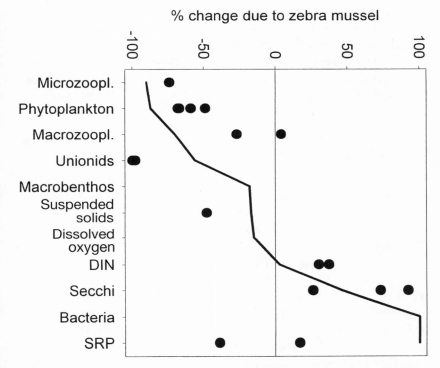

Figure 4.6. Ecosystem responses due to zebra mussels (*Dreissena polymorpha*) in the Hudson River (——) and in Lake Erie, Lake St. Clair, Saginaw Bay, and Oneida Lake (●). Modified from Strayer et al. (1999) and Caraco et al. (2000). DIN = dissolved inorganic nitrogen; Secchi = secchi disk; SRP = soluble reactive phosphorus.

1991), providing a longer-term perspective for evaluating the impact of this alien species. The effects of this invader on the structure and function of this large and complex freshwater ecosystem have been great, and in several instances unexpected (Figure 4.6). In particular, by 1993–96 populations of native mussels had decreased by approximately 60 percent, chlorophyll-*a* in the water-column decreased by approximately 85 percent, and dissolved oxygen decreased by approximately 15 percent (Caraco et al. 1997, 2000; Strayer et al. 1999). Similar ecosystem responses to zebra mussel invasion have been observed in Lake Erie, Lake St. Clair, Saginaw Bay, and Oneida Lake (see Figure 4.6; Strayer et al. 1999; Makarewicz, Bertram, and Lewis 2000). Normally, a sharp decline in dissolved oxygen, as was observed in the Hudson River, would be attributed to input of sewage or some other large source of allochthonous carbon. The oxygen decline in the Hudson River ecosystem, however, was especially surprising because it was not explained by these types of inputs (Caraco et al. 2000).

Peter Vitousek (1990) found that *Myrica faya,* an alien nitrogen-fixing plant introduced into Hawaii in the late 1800s, added large amounts of nitrogen to recent volcanic sites where nitrogen-fixing plants had not occurred before. Such fundamental changes in biogeochemical flux and cycling are important ecosystem-level effects of alien species invasions.

Value of Ecosystem Services

Recently much effort has been given to attempts at valuing the "services" provided by ecosystems for human and global life support systems, for instance, clean air, clean water, clean soil, and clean and nourishing food (Costanza et al. 1997; Daily et al. 1997; Barrett and Odum 2000; Carpenter and Turner 2000). Obviously this effort is extremely difficult, primarily because of the need to place "value" on components that different individuals, cultures, and ecosystems may assess differently. Moreover, these services are usually valued relative to human welfare, but "services" to other species are equally important if ecosystem "health" (structure and function) is to be sustained. For many years the Odums have argued for a common unit for valuation other than money (Odum 1975, 1996; Odum and Odum 2000).

A simple example of a critical ecosystem service is the provision of adequate supplies of clean water. During the past fifty years a liter of bottled water in the United States has become appreciably more expensive than a liter of bottled refined engine oil. Americans drink about forty-eight liters/person-year of bottled water (Hebert 1999). It is a sad commentary, in my opinion, that in our technologically advanced country we are reduced to drinking bottled water in the interest of good health (being fashionable also is heavily involved here, but there are serious concerns about the safety of public drinking water supplies across the United States). Nevertheless, the availability and sale of bottled water is only an anecdote about a very serious problem. On the basis of current information and projections (e.g., Francko and Wetzel 1983; Likens 1992; Population Action International Report 1994; Postel, Daily, and Ehrlich 1996; Gleick 1998; Vörösmarty et al. 2000), it is clear that availability of water throughout the globe will be *the* critical natural resource issue of the next decade or so. Obvious quality concerns include salinization, pathogens, toxic algal blooms, chemical pollution, eutrophication, and decreased species richness. Quantity concerns include development and application of new technologies, incongruent spatial dimensions of availability, and demand and transport, especially across political boundaries.

A simple but clear example of the role and value of ecosystem services in providing human support systems is shown by the New York City water supply system (see Chichilnsky and Heal 1998; Heal 1999). This system has been maintained and protected as remote, mostly forested watershed-ecosystems that provide clean drainage water, which is collected in reservoirs and transported via aqueducts to New York City. This ecosystem service has reliably provided high-quality water for a large urban population at relatively very low cost for decades. Currently, there is much concern about the potential compromise of this service because of air pollution inputs (Weathers et al. 2000; Lovett, Weathers, and Sobczak 2000) and exotic waterborne pathogens, such as *Giardia* sp. and *Cryptosporidium* sp. (NYCDEP 1993). If it were not for the watershed-ecosystems services, it has been estimated that the cost to filter the water supply for New York City would approach $9 billion, yet a filtration system would not sequester carbon, support biodiversity, or provide recreational opportunities (Chichilnsky and Heal 1998; Heal 1999).

FRONTIERS

Environmental change is more than just global warming or global climate change, which get most of the headlines (see, e.g., Vörösmarty et al. 2000). Clearly, human-accelerated environmental change includes as well problems such as loss of stratospheric ozone, loss of species, invasion of alien species, toxification of the biosphere, and changes in land use (Likens 1991). Even more important are the linkages and feedbacks among these various human-accelerated environmental changes, occurring simultaneously as well as with significant, incongruent legacies (Likens 1994).

We are rushing into the twenty-first century, and many important factors affecting ecosystem structure and function (e.g., size and impacts of the human population, carbon dioxide concentration in the atmosphere, amount of fertilizer and pesticide use, degradation of marine resources, landscape fragmentation, growth and number of urban agglomerations, rate of species loss) throughout the world are changing at an unprecedented rate. Are such rates of growth and utilization of resources, and the ecosystems that such changes affect, sustainable? Such questions will place great pressure and responsibility on ecosystem ecologists for answers. For example, human activity has approximately doubled the rate of nitrogen input into the terrestrial nitrogen cycle worldwide (Vitousek et al. 1997a). Because nitrogen is frequently a limiting nutrient in many aquatic and terrestrial ecosystems (Vitousek and Howarth 1991), this disturbance in the nitrogen cycle is a critical perturbation with many ecological ramifications and effects on interconnected human-accelerated environmental change issues.

Earth Day began on April 22, 1970, as a public movement, with much fanfare, excitement, and commitment to protect the Earth's thin and fragile environment. Yet we know now·that every square centimeter of the Earth's surface (even the most remote areas of Antarctica) has received and been degraded by human pollutants such as pesticides and toxic lead compounds.

For most people living in undeveloped areas of the world, survival depends directly on availability and use of natural resources. For most people in the most developed areas, however, the environment is part of some larger equation. Destruction is an inevitable consequence of people doing what they perceive to be "right." The fouling of rivers and the destruction of tribal territory to mine gold, the burning of jungle to raise beef cattle for fast-food hamburgers, the devastation of entire countrysides through industrial pollution in eastern Europe, the clear-cutting of old forests for export, the building of condominiums on filled wetlands—these are not ends in themselves. They are the by-products of economic development, job creation, the accumulation of wealth, and the quest for a better life. Unfortunately, what seems right for the short term may not be right for the long term. Quantitative understanding of ecosystem structure, function, and change with time offers hope for guiding better ecologic and economic management of such environmental problems and for guiding what may be right for sustaining environmental quality (see, e.g., Barrett and Odum 2000).

Ecosystem ecology is a relatively young science. The ecosystem approach is a powerful integrating tool for unraveling and simplifying the complexity of large areas. Thus, there are many exciting opportunities (e.g., increasing availability of data and information, major regional and global-scale changes ["experiments"] under way, longer-term financial support) and challenges (see Table 4.4) for unraveling the enormous complexity comprising diverse ecosystems and guiding solutions to environmental problems (see Pace and Groffman 1998). The present also is an exciting time as we learn from past breakthroughs in ecosystem science and look forward to new challenges and frontiers. We will need to develop new levels of understanding from all of Ecology to simplify complexity, so that suitable environments can be sustained on our planet for maintenance of diverse species, including humans (Likens 2001).

Probably never before have ecosystem ecologists needed to be as creative, innovative, proactive, and aggressive to meet the challenges of the next fifty years (Myers 1996; Lubchenco 1998; Ayensu et al. 1999). Likewise, never before has a holistic, comprehensive ecosystem approach (incorporating information from across the spectrum of Ecology to evolutionary biology to biogeochemistry) been more needed. I enthusiastically support the recent call for an assessment of the status and change in ecosystems throughout the world (Ayensu et al. 1999). Such an assessment is badly needed, especially in the face of the dramatic, often adverse, human-accelerated environmental changes worldwide, but it will be very difficult to do—more difficult than the Intergovernmental Panel on Climate Change assessment of global warming (IPCC 1990)—because ecosystems are so complex and diverse.

This is also a sobering time to study ecosystems because, as far as we know, it is the first time that one species has had such a dramatic impact on the environment of our planet. Because of its comprehensive nature, ecosystem science offers a hopeful approach for managing current and future large-scale problems in our search for environments to sustain us and other species into the twenty-first century and beyond.

Much of what I have called for and stressed in this chapter rests on the need for more integrated, comprehensive knowledge and understanding as a basis for management. This quest for knowledge is what I do and feel comfortable with as a scientist. It will not be enough, however, to bring about progressive action toward achieving specified ecosystem function and life support. Unless there is a crisis affecting people's welfare, the new knowledge and understanding will not be accepted and used quickly or widely. This reality is why I and others have called for a combination of biological-chemical-physical, socioeconomic, and cultural approaches toward integrated assessment and management of ecosystems. *These approaches must be integrated from the start to be effective.* Personally, I do not know clearly how to meet this difficult challenge, and the failures of the past, such as the National Science Foundation's 1970s initiative Interdisciplinary Research Relevant to Problems of Our Society, are sobering, but the urgency and increasing magnitude of the environmental problems we face demand nothing less than our best efforts. I also have made heavy use of words like *integration, complexity, sustainability*—however, none of this rhetoric will be fully successful unless we find ways to understand and clearly define these concepts and then to work together to implement solutions. Ecosystem ecology, with its holistic, comprehensive approach, offers a foundation for guiding these approaches and for providing solutions to currently widespread and rapidly developing environmental problems.

CHALLENGES FOR A NEW CENTURY

Some important ecosystem questions for the future (see also Table 4.4):

1. Can (should) ecosystems be modified by genetic engineering to enhance environmental quality and stability?

2. Can ecosystems be made to achieve specific functional goals (e.g., restoration)?
3. Can we build large and expanding urban agglomerations to function as net ecosystem producers rather than as heterotrophic ecosystems (e.g., techno-ecosystems; Naveh 1982)?
4. Can large landscapes consisting of a complex mosaic of land uses and social needs be managed for sustainable ecosystem services?
5. Can unexpected consequences or "surprises" be dealt with better or at least anticipated?
6. Can the relationships between biodiversity and ecosystem function be identified and utilized?

ACKNOWLEDGMENTS

Financial support was provided by the Institute of Ecosystem Studies and The Andrew W. Mellon Foundation. I acknowledge with appreciation the input of ideas and suggestions by E. Bernhardt, P. Cullen, M. Davis, J. Franklin, G. Glatzel, G. Harris, L. Hedin, R. Howarth, J. McCutchan, R. Stelzer, S. Tartowski, the scientific staff of the Institute of Ecosystem Studies (A. Berkowitz, C. Canham, J. Cole, S. Findlay, P. Groffman, K. Hogan, G. Lovett, M. Pace, R. Winchcombe, and especially N. Caraco, C. Jones, R. Ostfeld, S. Pickett, D. Strayer, J. Warner, and K. Weathers), and the technical support of D. Buso, D. Fargione, P. Likens, and J. Purrenhage. I also appreciate the thoughtful comments on the manuscript of G. Barrett and two anonymous reviewers.

LITERATURE CITED

Aber, J. D., C. T. Driscoll, C. A. Federer, R. Lathrop, G. M. Lovett, J. M. Melillo, P. Steudler, and J. Vogelmann. 1993. A strategy for the regional analysis of the effects of physical and chemical climate change on biogeochemical cycles in northeastern (U.S.) forests. *Ecol. Modelling* 67:37–47.

Ayensu, E., D. van R. Claasen, M. Collins, A. Dearing, L. Fresco, M. Gadgil, H. Gitay, G. Glaser, C. Juma, J. Krebs, R. Lenton, J. Lubchenco, J. McNeeley, H. Mooney, P. Pinstrup-Andersen, M. Ramos, P. Raven, W. Reid, C. Samper, J. Sarukhán, P. Schei, J. Galizia Tundisi, R. Watson, Xu Guanhua, and A. Zakri. 1999. International Ecosystem Assessment. *Science* 286:685–86.

Barrett, G. W. 1968. The effects of an acute insecticide stress on a semi-enclosed grassland ecosystem. *Ecology* 49:1019–35.

Barrett, G. W., and E. P. Odum. 2000. The twenty-first century: The world at carrying capacity. *BioScience* 50:363–68.

Basnet, K., G. E. Likens, F. N. Scatena, and A. E. Lugo. 1992. Hurricane Hugo: Damage to a tropical rain forest in Puerto Rico. *J. Tropical Ecol.* 8:47–56.

Batie, S. S. 1993. *Soil and water quality, an agenda for agriculture.* Washington, D.C.: National Academy Press.

Bax, N. J. 1999. Eradicating a dreissenid from Australia. *Dreissnena: The Digest of National Aquatic Nuisance Species Clearinghouse* 10:1–4.

Bormann, F. H., and G. E. Likens. 1967. Nutrient cycling. *Science* 155:424–29.

———. 1979. *Pattern and process in a forested ecosystem.* New York: Springer-Verlag.

Botkin, D. B., J. F. Janak, and J. R. Wallis. 1972. Rationale, limitations, and assumptions of a northeastern forest growth simulator. *IBM J. Res. Dev.* 16:101–16.

Bright, C. 1999. Invasive species: Pathogens of globalization. *Foreign Policy,* Fall, 50–64.

Burke, I. C., W. K. Lauenroth, and C. A. Wessman. 1998. Progress in understanding biogeochemical cycles at regional to global scales. In *Successes, limitations, and frontiers in ecosystem science,* edited by M. L. Pace and P. M. Groffman. New York: Springer- Verlag.

Caraco, N. F., J. J. Cole, S. E. G. Findlay, D. T. Fischer, G. G. Lampman, M. L. Pace, and D. L. Strayer. 2000. Dissolved oxygen declines in the Hudson River associated with the invasion of the zebra mussel *(Dreissena polymorpha). Environ. Sci. Technol.* 34:1204–10.

Caraco, N., J. J. Cole, P. A. Raymond, D. L. Strayer, M. L. Pace, S. E. G. Findlay, and D. T. Fischer. 1997. Zebra mussel invasion in a large, turbid river: Phytoplankton response to increased grazing. *Ecology* 78:588–602.

Carpenter, S. R. 1998. The need for large-scale experiments to assess and predict the response of ecosystems to perturbation. In *Successes, limitations, and frontiers in ecosystem science,* edited by M. L. Pace and P. M. Groffman. New York: Springer-Verlag.

Carpenter, S. R., D. E. Armstrong, D. E. Bennett, B. M. Kahn, K. J. Brasier, R. C. Lathrop, P. J. Nowak, and T. Reed. 2001. The ongoing experiment: Restoration of Lake Mendota. In *Lakes in the landscape: Long-term ecological research of north temperate lakes,* edited by J. J. Magnuson and T. K. Kratz. Cambridge: Oxford Univ. Press. In prep.

Carpenter, S. R., S. W. Chisholm, C. J. Krebs, D. W. Schindler, and R. F. Wright. 1995. Ecosystem experiments. *Science* 269:324–27.

Carpenter, S. R., and J. F. Kitchell, eds. 1993. *The trophic cascade in lakes.* London: Cambridge Univ. Press.

Carpenter, S. R., and M. Turner. 2000. Opening the black boxes: Ecosystem science and economic valuation. *Ecosystems* 3:1–3.

Carson, R. 1962. *Silent spring.* Boston: Houghton Mifflin.

Chichilnisky, G., and G. M. Heal. 1998. Economic returns from the biosphere. *Nature* 391:629–30.

Cole, J. J., G. M. Lovett, and S. E. G. Findlay, eds. 1991. *Comparative analyses of ecosystems: Patterns, mechanisms, and theories.* New York: Springer-Verlag.

Colwell, R. 2000. The role and scope for an integrated community of biologists. *BioScience* 50:199–202.

Costanza, R., R. d'Arge, R. de Groot, S. Farber, M. Grasso, B. Hannon, S. Naeem, K. Limburg, J. Paruelo, and R. V. O'Neill. 1997. The value of the world's ecosystem services and natural capital. *Nature* 387:253–60.

Crutzen, P. J. 1971. Ozone production rates in an oxygen-hydrogen-nitrogen oxide atmosphere. *J. Geophys. Res.* 76:7311–27.

Daily, G. C., P. R. Ehrlich, L. H. Goulder, J. Lubchenco, P. A. Matson, H. A. Mooney, S. H. Schneider, G. M. Woodwell, and D. Tilman. 1997. Ecosystem services: Benefits supplied to human societies by natural ecosystems. *Issues in Ecology* 2:1–16.

Easterbrook, G. 1995. *A moment on the Earth: The coming age of environmental optimism.* New York: Penguin Books.

Elser, J. J., D. R. Dobberfuhl, N. A. MacKay, and J. H. Schampel. 1996. Organism size, life history, and N:P stoichiometry: Towards a unified view of cellular and ecosystem processes. *BioScience* 46:674–84.

Elser, J. J., and J. Urabe. 1999. The stoichiometry of consumer-driven nutrient recycling: Theory, observations, and consequences. *Ecology* 80:735–51.

Elton, C. 1939. *Animal ecology.* New York: Macmillan. Reprint of 1927. London: Sidgwick and Jackson.

Evans, F. C. 1956. Ecosystem as the basic unit in ecology. *Science* 123:1127–28.

Francko, D. A., and R. G. Wetzel. 1983. *To quench our thirst: Present and future freshwater resources of the United States.* Ann Arbor: Univ. of Michigan Press.

Galloway, J. N., H. Levy II, and P. S. Kasibhatla. 1994. Year 2020: Consequences of population growth and development on the deposition of oxidized nitrogen. *Ambio* 23:120–23.

Galloway, J. N., G. E. Likens, W. C. Keene, and J. M. Miller. 1982. The composition of precipitation in remote areas of the world. *J. Geophys. Res.* 87:8771–86.

Gleick, P. H. 1998. *The world's water.* Washington, D.C.: Island Press.

Golley, F. B. 1993. *A history of the ecosystem concept in ecology.* New Haven: Yale Univ. Press.

Gosz, J. R., R. T. Holmes, G. E. Likens, and F. H. Bormann. 1978. The flow of energy in a forest ecosystem. *Sci. Amer.* 238:92–102.

Hagen, J. B. 1992. *An entangled bank: The origins of ecosystem ecology.* New Brunswick, N.J.: Rutgers Univ. Press.

Hasler, A. D., ed. 1975. *Proceedings of the INTECOL symposium on coupling of land and water systems, 1971.* New York: Springer-Verlag.

Hasler, A. D., O. M. Brynildson, and W. T. Helm. 1951. Improving conditions for fish in brown-water lakes by alkalization. *J. Wildl. Manage.* 15:347–52.

Heal, G. M. 1999. Valuing ecosystem services. In *Money, economics, and finance.* Columbia Univ., Columbia Business School.

Hebert, H. J. 1999. Study raises questions about just how pure bottled water is. *Ithaca Journal*, Mar. 31.

Hedin, L. O., J. J. Armesto, and A. H. Johnson. 1995. Patterns of nutrient loss from unpolluted, old-growth temperate forests: Evaluation of biogeochemical theory. *Ecology* 76:493–509.

Holden, C. 2000. From ballast to bouillabaisse. *Science* 289:241.

Hrbácek, J., M. Dvoráková, M. Korínek, and L. Procházková. 1961. Demonstration of the effect of the fish stock on the species composition of zooplankton and the intensity of metabolism of the whole plankton association. *Verh. der Internat. Verein. für Theor. Ang. Limnol.* 14:192–95.

Hutchinson, G. E. 1950. Survey of contemporary knowledge of biogeochemistry: III. The biogeochemistry of vertebrate excretion. *Bull. Amer. Mus. Nat. Hist.* 96.

———. 1957. *A treatise on limnology.* Vol. 1 of *Geography, Physics, and Chemistry.* New York: John Wiley.

IPCC. 1990. *Climate change: The IPCC scientific assessment.* Cambridge: Cambridge Univ. Press.

Jones, C. G., and J. H. Lawton, eds. 1995. *Linking species and ecosystems.* New York: Chapman & Hall.

Jones, C. G., J. H. Lawton, and M. Shachak. 1994. Organisms as ecosystem engineers. *Oikos* 69:373–86.

Jones, C. G., R. S. Ostfeld, M. P. Richard, E. M. Schauber, and J. O. Wolff. 1998. Chain reactions linking acorns, gypsy moth outbreaks, and Lyme-disease risk. *Science* 279:1023–26.

Jones, C. G., and M. Shachak. 1990. Fertilization of the desert soil by rock-eating snails. *Nature* 346:839–41.

Juday, C. 1940. The annual energy budget of an inland lake. *Ecology* 21:438–50.

Kovda, V. A. 1975. *Biogeochemical cycles in nature and their study.* Moscow: Publishing Hause Hauka.

Lauenroth, W. K., C. D. Canham, A. P. Kinzig, K. A. Poiani, W. M. Kemp, and S. W. Running. 1998. Simulation modeling in ecosystem science. In *Successes, limitations, and frontiers in ecosystem science,* edited by M. L. Pace and P. M. Groffman. New York: Springer- Verlag.

Leopold, A. 1939. A biotic view of land. In *The river of the Mother of God,* edited by S. L. Flader and J. Baird Callicott. Madison: Univ. of Wisconsin Press.

Lewis, G. P., and G. E. Likens. 2001. Potential contribution of insect defoliation to elevated nitrate loss from a northern hardwood forest. In prep.

Likens, G. E. 1985. An experimental approach for the study of ecosystems. *J. Ecol.* 73:381–96.

———. 1991. Human-accelerated environmental change. *BioScience* 41:130.

———. 1992. *The ecosystem approach: Its use and abuse.* Vol. 3 of *Excellence in ecology.* Oldendorf/Luhe, Germany: Ecology Institute.

———. 1994. Human-accelerated environmental change: An ecologist's view. 1994 Australia Prize winner presentation. Perth: Murdoch Univ.

———. 1998. Limitations to intellectual progress in ecosystem science. In *Successes, limitations, and frontiers in ecosystem science,* edited by M. L. Pace and P. M. Groffman. New York: Springer-Verlag.

———. 2001. Eugene Odum, the ecosystem approach, and the future. In *Holistic science: The evolution of the Georgia Institute of Ecology (1940–2000),* edited by G. W. Barrett and T. L. Barrett. Newark, N.J.: Harwood Acad. Publ. In press.

Likens, G. E., ed. 1989. *Long-term studies in ecology: Approaches and alternatives.* New York: Springer-Verlag.

Likens, G. E., and F. H. Bormann. 1974a. Acid rain: A serious regional environmental problem. *Science* 184:1176–79.

———. 1974b. Linkages between terrestrial and aquatic ecosystems. *BioScience* 24:447–56.

———. 1985. An ecosystem approach. In *An ecosystem approach to aquatic ecology: Mirror Lake and its environment,* edited by G. E. Likens. New York: Springer-Verlag.

———. 1995. *Biogeochemistry of a forested ecosystem.* 2d ed. New York: Springer-Verlag.

Likens, G. E., F. H. Bormann, and N. M. Johnson. 1969. Nitrification: Importance to nutrient losses from a cutover forested ecosystem. *Science* 163:1205–6.

———. 1972. Acid rain. *Environment* 14:33–40.

Likens, G. E., F. H. Bormann, N. M. Johnson, D. W. Fisher, and R. S. Pierce. 1970. Effects of forest cutting and herbicide treatment on nutrient budgets in the Hubbard Brook watershed-ecosystem. *Ecol. Monogr.* 40:23–47.

Likens, G. E., F. H. Bormann, R. S. Pierce, J. S. Eaton, and N. M. Johnson. 1977. *Biogeochemistry of a forested ecosystem.* New York: Springer-Verlag.

Likens, G. E., C. T. Driscoll, D. C. Buso, T. G. Siccama, C. E. Johnson, G. M. Lovett, T. J. Fahey, W. A. Reiners, D. F. Ryan, C. W. Martin, and S. W. Bailey. 1998. The biogeochemistry of calcium at Hubbard Brook. *Biogeochemistry* 41:89–173.

Lindeman, R. L. 1942. The trophic-dynamic aspect of ecology. *Ecology* 23:399–418.

Lovett, G. M., K. C. Weathers, and W. Sobczak. 2000. Nitrogen saturation and retention in forested watersheds of the Catskill Mountains, N.Y. *Ecol. Appl.* 10:73–84.

Lubchenco, J. 1998. Entering the century of the environment: A new social contract with science. *Science* 279:491–97.

McDonnell, M. J., and S. T. A. Pickett, eds. 1993. *Humans as components of ecosystems: The ecology of subtle human effects and populated areas.* New York: Springer-Verlag.

McIntosh, R. P. 1985. *The background of ecology: Concept and theory.* Cambridge: Cambridge Univ. Press.

Makarewicz, J. C., P. Bertram, and T. W. Lewis. 2000. Chemistry of the offshore surface waters of Lake Erie: Pre- and post-Dreissena introduction (1983–1993). *J. Great Lakes Res.* 26:82–93.

Mallin, M. A. 2000. Impacts of industrial animal production on rivers and estuaries. *Amer. Sci.* 88:26–37.

Martin, M. E., and J. D. Aber. 1997. Estimation of forest canopy lignin and nitrogen concentration and ecosystem processes by high spectral resolution remote sensing. *Ecol. Applications* 7:431–43.

Martin, M.E., S. D. Newman, J. D. Aber, and R. G. Congalton. 1998. Determining forest species composition using high spectral resolution remote sensing data. *Remote Sensing of the Environment* 65:249–54.

Matson, P. A., and S. L. Ustin. 1991. Special feature: The future of remote sensing in ecological studies. *Ecology* 76:1917.

Mitchell, M. J., C. T. Driscoll, J. S. Kahl, G. E. Likens, P. S. Murdoch, and L. H. Pardo. 1996. Climatic control of nitrate loss from forested watersheds in the Northeast United States. *Environ. Sci. Technol.* 30:2609–12.

Molina, M. J., and F. S. Rowland. 1974. Stratospheric sink for chlorofluoromethanes: Chlorine atom-catalysed destruction of ozone. *Nature* 249:810–12.

Munn, T. E., A. Whyte, and P. Timmerman. 1999. Emerging environmental issues: A global perspective of SCOPE. *Ambio* 28:464–71.

Myers, N. 1980. *Conversion of tropical moist forests*. Washington, D.C.: National Academy Press.

———. 1996. Development, environment, and health: What else we should know? *Environ. and Dev. Econ.* 1:367–71.

Naveh, Z. 1982. Landscape ecology as an emerging branch of human ecosystem science. *Adv. Ecol. Res.* 12:189–237.

New York City Department of Environmental Protection (NYCDEP). 1993. *New York City drinking water quality control, 1992 watershed annual report*. New York: NYCDEP.

Odèn, S. 1968. *The acidification of air and precipitation and its consequences on the natural environment*. Swedish National Science Research Council, Ecology Committee, Bull. 1.

Odum, E. P. 1959. *Fundamentals of ecology*. 2d ed. Philadelphia: Saunders.

———. 1964. The new ecology. *BioScience* 14:14–16.

———. 1969. The strategy of ecosystem development. *Science* 164:262–70.

Odum, H. T. 1956. Primary production in flowing waters. *Limnol. Oceanogr.* 1:102–17.

———. 1957. Trophic structure and productivity of Silver Springs, Florida. *Ecol. Monogr.* 27:55–112.

———. 1960. Ecological potential and analogue circuits for the ecosystem. *Amer. Sci.* 48:1–8.

———. 1975. Energy quality and the carrying capacity of the Earth. *Tropical Ecol.* 16:1–8.

———. 1996. *Environmental accounting: EMERGY and environmental decision making*. New York: John Wiley.

Odum, H. T., and E. P. Odum. 2000. The energetic basis for valuation of ecosystem services. *Ecosystems* 3:21–23.

Office of Technology Assessment. 1993. *Harmful nonindigenous species in the United States*. Washington, D.C.: OTA.

Ostfeld, R. S. 1997. The ecology of Lyme-disease risk. *Amer. Sci.* 85:338–46.

Ostfeld, R. S., and C. G. Jones. 1999. Peril in the understory. *Audubon*, July–Aug., 74–82.

Ostfeld, R. S., C. G. Jones, and J. O. Wolff. 1996. Of mice and mast: Ecological connections in eastern deciduous forests. *BioScience* 46:323–30.

Ostfeld, R., F. Keesing, C. G. Jones, C. D. Canham, and G. M. Lovett. 1998. Integrative ecology and the dynamics of species in oak forests. *Integrative Biology* 1:178–85.

Pace, M. L., and P. M. Groffman, eds. 1998. *Successes, limitations, and frontiers in ecosystem science*. New York: Springer-Verlag.

Pickett, S. T. A., R. S. Ostfeld, M. Shachak, and G. E. Likens, eds. 1997. *The ecological basis of conservation: Heterogeneity, ecosystems, and biodiversity*. New York: Chapman & Hall.

Pickett, S. T. A., and P. S. White, eds. 1985. *The ecology of natural disturbance and patch dynamics*. New York: Academic Press.

Pimental, D., L. Lach, R. Zuniga, and D. Morrison. 2000. Environmental and economic costs of nonindigenous species in the United States. *BioScience* 50:53–65.

Population Action International Report. 1994. *Sustaining water: An update*. Revised data for the Population Action International Report. Washington, D.C.: Population and Environment Program.

Postel, S. L., G. C. Daily, and P. R. Ehrlich. 1996. Human appropriation of renewable fresh water. *Science* 271:785–88.

Power, M. E., D. Tilman, J. A. Estes, B. A. Menge, W. J. Bond, L. S. Mills, G. Daily, J. C. Castilla, J. Lubchenco, and R. T. Paine. 1996. Challenges in the quest for keystones. *BioScience* 46:609–20.

Redfield, A. C. 1958. The biological control of chemical factors in the environment. *Amer. Sci.* 46:205–21.

Rees, W. E., and M. Wackernagel. 1994. Ecological footprints and appropriated carrying capacity: Measuring the natural capital requirements of the human economy. In *Investing in natural capital*, edited by A. M. Jansson, M. Hammer, C. Folke, and R. Costanza. Washington, D.C.: Island Press.

Reiners, W. A. 1986. Complementary models for ecosystems. *Amer. Natur.* 127:60–73.

Richards, J. F. 1990. Land Transformation. In *The Earth as transformed by human action*, edited by B. L. Turner II, W. C. Clark, R. W. Kates, J. F. Richards, J. T. Mathews, and W. B. Meyer. Cambridge: Cambridge Univ. Press.

Rowland, F. S., and M. J. Molina. 1975. Chlorofluoromethanes in the environment. *Rev. Geophys. Space Phys.* 13:1–35.

Schindler, D. W. 1973. Experimental approaches to limnology: An overview. *J. Fish. Res. Bd. Canada* 30:1409–13.

———. 1977. The evolution of phosphorus limitation in lakes. *Science* 195:260–62.

Schindler, D. W., K. H. Mills, D. F. Malley, D. L. Findlay, J. A. Shearer, I. J. Davies, M. A. Turner, G. A. Linsey, and D. R. Cruikshank. 1985. Long-term ecosystem stress: The effects of years of experimental acidification on a small lake. *Science* 228:1395–1401.

Schultz, V., and A. W. Klement, Jr., eds. 1963. *Radioecology*. New York: Reinhold and Washington, D.C.: AIBS.

Shachak, M., and C. G. Jones. 1995. Ecological flow chains and ecological systems: Concepts for linking species and ecosystem perspectives. In *Linking species and ecosystems*, edited by C. G. Jones and J. H. Lawton. New York: Chapman & Hall.

Shachak, M., C. G. Jones, and S. Brand. 1995. The role of animals in an arid ecosystem: Snails and isopods as controllers of soil formation, erosion, and desalinization. *Adv. Geo. Ecol.* 28:37–50.

Shapiro, J. 1979. The need for more biology in lake restoration. In *Lake restoration*, U.S. EPA 440/5-79-001. Washington, D.C.: EPA.

Slobodkin, L. B. 1962. Energy in animal ecology. In *Advances in ecological research 12*, edited by J. B. Cragg. London: Academic Press.

Smith, V. H. 1998. Cultural eutrophication of inland, estuarine, and coastal waters. In *Successes, limitations, and frontiers in ecosystem science*, edited by M. L. Pace and P. M. Groffman. New York: Springer-Verlag.

Stearns, F., and T. Montag, eds. 1974. *The urban ecosystems: A holistic approach*. Stroudsburg, Pa.: Dowden, Hutchinson and Ross.

Sterner, R. W., J. J. Elser, E. J. Fee, S. J. Guildford, and T. H. Chrzanowski. 1997. The light:nutrient ratio in lakes: The balance of energy and materials affects ecosystem structure and process. *Amer. Natur.* 150:663–84.

Strayer, D. L., N. F. Caraco, J. J. Cole, S. Findlay, and M. L. Pace. 1999. Transformation of freshwater ecosystems by bivalves: A case study of zebra mussels in the Hudson River. *BioScience* 49:19–27.

Strayer, D. L., J. Powell, P. Ambrose, L. C. Smith, M. L. Pace, and D. T. Fischer. 1996. Arrival, spread and early dynamics of a zebra mussel (*Dreissena polymorpha*) population in the Hudson River estuary. *Can. J. Fish. Aquatic Sci.* 53:1143–49.

Stumm, W., ed. 1977. *Global chemical cycles and their alterations by man*. Berlin: Dahlem Konferenzen.

Swank, W. T., and D. A. Crossley, Jr. 1988. *Forest hydrology and ecology at Coweeta*. New York: Springer-Verlag.

Tansley, A. G. 1935. The use and abuse of vegetational concepts and terms. *Ecology* 16:284–307.

Turner, B. L., II, W. C. Clark, R. W. Kates, J. F. Richards, J. T. Mathews, and W. B. Meyer, eds. 1990. *The Earth as transformed by human action*. New York: Cambridge Univ. Press.

Van Dyne, G. M. 1966. *Ecosystems, systems ecology, and systems ecologists*. ORNL-3957. Oak Ridge, Tenn.: ORNL.

Vernadsky, W. I. 1944. Problems in biogeochemistry, II. *Trans. Conn. Acad. Arts Sci.* 35:493–94.

———. 1945. The biosphere and the noösphere. *Amer. Sci.* 33:1–12.

Vitousek, P. M. 1990. Biological invasions and ecosystem processes: Towards an integration of population biology and ecosystem studies. *Oikos* 57:7–13.

———. 1994. Beyond global warming: Ecology and global change. *Ecology* 75:1861–76.

Vitousek, P. M., J. D. Aber, R. W. Howarth, G. E. Likens, P. A. Matson, D. W. Schindler, W. H. Schlesinger, and D. G. Tilman. 1997a. Human alteration of the global nitrogen cycle: Sources and consequences. *Ecol. Appl.* 7:737–50.

Vitousek, P. M., and R. W. Howarth. 1991. Nitrogen limitation on land and in the sea: How can it occur? *Biogeochemistry* 13:87–115.

Vitousek, P. M., H. A. Mooney, J. Lubchenco, and J. M. Melillo. 1997b. Human domination of Earth's ecosystems. *Science* 277:494–99.

Vollenweider, R. A. 1968. *Scientific fundamentals of lake and stream eutrophication, with particular reference to phosphorus and nitrogen as eutrophication factors.* Technical Report DAS/DSI/68.27. Paris: OECD.

Vörösmarty, C. J., P. Green, J. Salisbury, and R. B. Lammers. 2000. Global water resources: Vulnerability from climate change and population growth. *Science* 289:284–88.

Watt, K. E. F., ed. 1966. *Systems analysis in ecology.* New York: Academy Press.

Weathers, K. C., G. M. Lovett, G. E. Likens, and R. Lathrop. 2000. The effect of landscape features on deposition to Hunter Mountain, Catskill Mountains, N.Y. *Ecol. Appl.* 10:528–40.

Wessman, C. A., and G. P. Asner. 1998. Ecosystems and problems of measurement at large spatial scales. In *Successes, limitations, and frontiers in ecosystem science,* edited by M. L. Pace and P. M. Groffman. New York: Springer-Verlag.

5

Behavior, Ecology, and Evolution

GORDON H. ORIANS

This chapter attempts to assess the current status of biological knowledge and
to highlight some major challenges and opportunities facing investigators in the
field of animal behavior. The editors of this book have asked us to be bold in
our speculations. As I rise to that challenge I am aware on the one hand of Mark
Twain's observation that "there is something fascinating about science, one gets
such wholesale returns of conjecture out of such trifling investments of fact."
On the other hand, failure to conceive of imaginative possibilities may cause us
to miss major insights.

To identify major challenges in current and evolutionary relationships between behavior and ecology requires that I be selective because the field is so
vast and many important challenges exist. I will therefore concentrate on four
challenges we face as we attempt to achieve a better synthesis of concepts that
find their foci at different levels of biological organization, from single cells to
ecosystems. First, I will consider challenges in understanding the development
of a functioning adult from a single-celled zygote, viewed specifically from a
behavioral perspective. Second, I will discuss how the structure and functioning of ecological systems is influenced by the behavior of their constituent individuals. Third, I will address how behavior influences evolutionary processes.
Finally, I will explore how we can determine the course of events leading to past
and current behavior patterns.

The setting for this chapter was concisely stated by G. Evelyn Hutchinson
(1965) with his metaphor "the ecological theater and the evolutionary play."
The actors in the evolutionary play are, of course, engaged in a variety of behaviors. The majority of the properties of ecological systems are the results of

the behavior of myriad individuals interacting with one another and their physical environment. Indeed, in a fundamental sense ecology is the science that attempts to determine both the consequences of behavior for the structure and functioning of ecological systems and the influence of those complex systems on the behavioral characteristics that evolve.

Although what I have just stated may seem trivially obvious to us today, the study of animal behavior during much of the past century is characterized by a failure to pay adequate attention to the ecological theater. Part of the reason is that the goal of much behavioral research was to understand human behavior. Animals were employed as surrogates because they could be manipulated in ways that were prohibited with humans. The view that dominated thinking in the social sciences for many decades was that, although humans had clearly evolved from primate ancestors, human behavior was no longer influenced by human genetics. The causes of general patterns and the rich variability in human behavior were to be sought in studies of the experiences of individuals as they developed and matured in the complex social matrices that characterize human societies.

This conceptual framework, in which ecology had little or no role, encouraged an extensive search for general laws of learning. That search failed in large part because it did not consider the rich and varied roles of learning in the lives of organisms. It was not until the 1960s that experiments were performed to elucidate specific features of learning in the context of responses to specific problems faced by organisms. Illustrative of the new innovations are the experiments of J. García and R. A. Koelling (1966) on avoidance learning in rats. In those experiments drinking rats were confronted with a stimulus that contained both a taste element and an audiovisual element. After a drinking session half of the rats were made sick by means of X radiation; the other half received an electrical shock. Later the rats were made thirsty and offered an opportunity to drink in the presence of one of the two experimental stimuli. Those animals that had been made sick avoided water flavored with the taste they had previously experienced. These animals did not avoid water when their drinking was accompanied by the audiovisual stimulus. In contrast, the animals that had received an electrical shock avoided drinking the water when it was accompanied by the audiovisual stimulus, but they did not avoid water that was flavored with the taste they had previously experienced.

García and Koelling offered this interpretation of their data: "Natural selection may have favored mechanisms which associate gustatory and olfactory cues with internal discomfort, since the chemical receptors sample the materials soon to be incorporated into the internal environment" (1966, p. 124). Thus, they interpreted their results in terms of the consequences of the mechanisms and the actions they produced for the survival of the individuals, and they recognized that the patterns required some genetic basis.

It is difficult today, when no serious student of animal behavior questions that genetic factors influence the complex patterns of learning (Seligman 1970), to imagine how revolutionary these results and interpretations were thirty-five years ago. We have come a long way during that brief interval. It is interesting that the only resistance to this view persisting in the scientific community concerns the influence of genetics on human behavior.

Ecologists, in turn, have had an ambivalent relationship with the field of animal behavior. Behavioral ecology has always been an important component of ecology, but population ecologists have made relatively little use of the rich results of behavioral ecological research (however, see Sutherland and Parker 1992; Fryxell and Lundberg 1998; Lima 1998). And as ecologists increasingly directed their attention to patterns of functioning of complex ecosystems, their interest in data on animal and plant behavior declined even further. Papers dealing with behavioral ecology have occupied a decreasing proportion of the programs of the annual meetings of the Ecological Society of America. Behavioral ecologists have, in turn, formed their own societies and attend their own meetings.

Part of the reason for the decline of behavioral ecology in ecological studies stems from the difficulty of incorporating behavioral data into models of ecosystem functioning. Models of complex systems must sacrifice details to render tractable the theoretical and simulation analysis of the major features of the systems. The behavior of individuals (and even of species) has typically been the element sacrificed. This is partly because of a lack of the necessary species-specific information, but it also reflects the fact that behavioral ecologists have, until recently, given little thought to the broader ecological consequences of the decision rules they have developed to describe the behavior of individuals.

In general terms then, a major challenge confronting investigators in the fields of behavior, ecology, and evolution is to find more powerful ways of integrating studies of the ecological theater and the actors that constitute the evolutionary play. Before addressing this challenge, I will present the conceptual framework for the study of behavior that I will use to identify the multilevel challenges to our understanding of behavior.

A CONCEPTUAL FRAMEWORK FOR THE STUDY OF BEHAVIOR

Early each spring thousands of monarch butterflies that have spent the winter high in the mountains west of Mexico City leave their wintering grounds and migrate north to the southern United States, where they mate, lay eggs, and die. Members of the next generation of butterflies fly farther north, lay eggs, and die.

Survivors of the next generation fly south to the mountains of Mexico, where they will spend the winter. Individuals performing these migratory behaviors are three generations removed from the individuals that last accomplished those acts (Brower 1977). How do they know what to do?

As pointed out many years ago by Niko Tinbergen (1951) and Ernst Mayr (1960), answers to questions about behavior fall into two major categories. One concerns the proximate causes of behavior, that is the genetic-developmental mechanisms that underlie behavior and the sensory-motor mechanisms (nervous systems for detection of environmental stimuli, hormonal systems for adjusting responsiveness to environmental stimuli, immunological systems for distinguishing self from nonself, skeletal-motor systems for carrying out responses) by means of which behavior actually happens. The other category deals with ultimate causes of behavior. These include the historical pathways leading to a current behavior (the events occurring over evolution from the origin of the trait to the present) and the selective pressures shaping the history of a behavioral trait (past and current contributions of the behavior to fitness).

Because the evolution of life is an ongoing process that began about 4 billion years ago, biology, like the mental world of Ebenezer Scrooge, is replete with ghosts. There are ghosts of predators past, ghosts of parasites past, ghosts of competitors past, ghosts of conspecifics past, and ghosts of meteors, volcanic eruptions, hurricanes, and droughts past. Some ghosts are carryovers from ancient events; others come from recent events. Identifying these evolutionary ghosts, their origins, and their current influences is difficult. A major challenge in the study of relationships between behavior and ecology is how best to identify, characterize, and interpret ghosts.

The environment provides information to an organism, but for that information to become knowledge that can guide adaptive responses, an organism must establish a relationship between the information and possible responses to it. Organisms do this by forming internal "models" of the information. Such models have characterized life from its origins to the present day. The complexity of both the models and the amount and kinds of information organisms can process has increased in many, but not all, organismal lineages.

The process by which internal models of the world have evolved appears to be Bayesian; that is, nervous systems evolving under the influence of natural selection have developed internal models that embody a priori expectations of the state of the world (Olsson and Holmgren 1998). Organisms judge the significance of incoming information using these a priori expectations and respond accordingly. Natural selection acts as the ongoing judge of the efficacy of those responses. Needless to say, none of this requires (but it does not preclude) cognitive awareness on the part of individuals. We need to assume only that animals act *as if* their responses have been molded through evolution by a Bayesian process. Given that natural selection has resulted in organisms that respond to their environments using a priori expectations, the increasing use among scientists of Bayesian statistics may be highly appropriate.

The knowledge organisms possess is bounded by the limits of the quality and quantity of information provided by sensors and by the capacity of the computing system behind them (Marchetti 1998). Over the nearly 4 billion years of organic evolution sensors have evolved impressive capacities. The chemical sensors of organisms, which have been evolving ever since the origin of life on Earth, have been perfected in some species to the limits imposed by quantum mechanics. That is, some sensors can detect a single molecule. Photoreceptors, which probably began with light-sensitive spots in bacteria, have not reached quantum limits. The most visually acute eyes, perhaps those of eagles, are impressive, but they do not approach the capacity of modern telescopes. Sound receptors are also impressive, but, because of the intrinsic properties of sound waves, with the exception of bat sonar systems, they generally provide more precise information about what than about where. It is interesting that electromagnetic radar sensory systems apparently have not evolved, even though production of electricity is well within the capacity of living systems (Marchetti 1998).

The limits of knowledge possessed by organisms have evolved in response to the costs of receiving and maintaining knowledge and the benefits the knowledge confers. Organisms have evolved their specific characteristics because of the costs and benefits they have conferred and imposed on their bearers (Goldsmith 1990).

A cost-benefit approach to knowledge is aided by using a framework whose components are genetic, developmental, adult learning, and sociocultural learning (Plotkin 1988). Processes in these four components differ markedly in their sensitivity to events occurring at different frequencies. The rate at which genetic changes can lead to adaptations is restricted temporally by the generation time of organisms. It is sensitive only to events of low frequency relative to generation time; responses cannot be made to events occurring at higher frequencies. To the extent that higher-frequency events influence the fitness of organisms, they become selective factors favoring the evolution of mechanisms in the other components. Developmental and adult learning processes can process and respond to information arriving at much higher frequencies, unconstrained by the generation times of organisms. And, of course, sociocultural learning can potentially proceed at an even more rapid rate, although, as human cultural conservatism amply demonstrates, it does not necessarily do so.

All four processes are profoundly influenced by genetic factors. During development the many possible trajectories that actually unfold are functions of both an organism's genotype and the environment it encounters. Similarly, although learning clearly is powerfully influenced by environmental events, capacities to learn and propensities to learn different things are strongly influenced by the genotypes of organisms, a phenomenon generally referred to as genetically programmed learning (Seligman 1970; Pulliam and Dunford 1980).

Although the genetic system offers the slowest rate of information acquisition, it has the lowest rate of error generation, and the most reliable temporal storage, and it is apparently metabolically cheap. The other processes operate faster, but they are more prone to errors of sampling and storage. The immunological system evolved to make a fundamental distinction between self and nonself. It does so by producing a diverse array of solutions layered onto arriving problems with the hope that one of them will fit. Although they originally may have evolved as parts of defenses against disease-causing organisms, the major histocompatibility complex (MHC) genes subsequently evolved to be components of learned kin recognition in some animals.

Because the MHC genes may have as many as fifty alleles per gene (in humans and house mice), individuals in a population differ in MHC genotype. Experiments with house mice have demonstrated that these alleles influence an individual's odor. Mice in inbred strains can distinguish among individuals that differ only in their MHC genes, and these mice, as well as mice in seminatural conditions, prefer to mate with individuals whose MHC genes differ from their own (Potts, Manning, and Wakeland 1991). The function of these discriminations may be to avoid inbreeding, to enhance immunocompetence, or to provide a rare-allele advantage against rapidly evolving parasites (Penn and Potts 1999).

The nervous system can respond to new information at a high rate. It has a high potential rate of transmission errors but a low rate of rigidity errors. And it is very expensive. The more information that must be stored at high rates, the more complex and expensive the central nervous system must be. The human central nervous system constitutes only 2 percent of body weight, but it consumes about 20 percent of metabolic energy.

Cost-benefit accounting in biology is done in fitness units. The costs of the knowledge system include the energy content of the code (very small), the energy invested in building and maintaining the knowledge acquisition and storage system, the time cost of acquiring the information, and the risks associated with information acquisition. Even though we are organisms with impressive learning capabilities, we often underestimate the costs of learning. Acquiring information can be risky. Curiosity, we are told, sometimes kills cats!

Substantial opportunity costs are associated with knowledge systems because time and energy invested in them are not available for other uses. For example, learning to drive a car requires concentration on the task, making it difficult to carry on a conversation or attend to anything else during the learning process. But when driving is learned and has become a "habit," we no longer need to "think about it." This fact has long been appreciated. Thus, in 1911 Alfred North Whitehead wrote, "It is a profoundly erroneous truism, repeated by all copybooks and by eminent people when they are making speeches, that we should cultivate the habit of thinking of what we are doing. The precise opposite is the case. Civilization advances by extending the number of operations which we can perform without thinking about them."

The benefits of investments in knowledge acquisition and storage systems are better responses to environmental information. Information varies greatly in importance and the time frame over which responses to it must be made if actions are to enhance fitness. Professors are often accused, with some justification, of committing the "academic fallacy," that is, of treating all information as equally important. At least the contents of our examinations often appear to suggest failure to discriminate between trivial and important information. Natural selection consistently avoids this error.

Forms of benefit curves vary greatly depending on the type of information and the problems to be solved. A common form is sigmoid; that is, a little knowledge is not worth much. In fact, a little knowledge may be dangerous! However, the value of additional knowledge often increases rapidly until an asymptote is reached. Such a relationship might pertain where a bit of knowledge is more confusing than clarifying or where opportunity costs are very high. A declining exponential benefit curve might exist where an environmental mean is being estimated by successive measurements and where risks associated with acquiring information are low. A major challenge in the study of behavior is to determine the forms of cost-benefit curves for a rich array of problem-solving activities.

If the forms of cost and benefit curves can be specified, cost-benefit models can be used to predict equilibrium solutions to the amount of knowledge an organism should possess. They can be applied to help us understand why very few organisms have chemosensory systems that approach quantum limits and why few eyes match those of eagles. This is easy to state in principle, but two challenging tasks remain. One is to infer, test, and modify as necessary our assessments of the curves in relation to specific problems organisms must solve. The other is to incorporate genetics and maternal or other parental effects into cost-benefit models, most of which are strictly phenotypic.

I now turn to the four major interlevel challenges.

FROM THE ZYGOTE TO THE FUNCTIONING ADULT: HOW DO ORGANISMS DO THE "RIGHT THING"?

Individual organisms choose habitats in which to forage and to breed, types of food to eat, where to seek shelter, how to avoid predators and parasites, and individuals with which to share genes and produce offspring. And because most of these behaviors are temporally mutually exclusive, they must decide when to do these things. Typically an individual cannot court and feed at the same time. Trade-offs characterize all decision making, and natural selection is the agent that arbitrates trade-offs.

Investigations of potential trade-offs are often aided by casting problems in an optimization framework (Maynard Smith 1978; Stephens and Krebs 1986; Mangel and Clark 1988). The typical approach is to identify appropriate biological limitations on an animal's behavior and incorporate them into an optimization model that predicts the strategy that maximizes some objective (energy maximization per unit time foraging, balancing a trade-off between predation risk and foraging, hedging against longer-term fluctuations in food availability). Empirical tests are then used to compare performances with the predicted optimal strategy.

To illustrate, let us consider a red-winged blackbird. Our hypothetical blackbird is foraging and has just encountered a potential food item. An appropriate response to the item requires the bird to identify it, assess its nutrient value, the energetic cost of pursuing it, the time to capture and consume it, and to assess the possible presence of predators and competitors. The bird's response is also made in the context of its current hunger level and the abundance of other food items in the environment and their characteristics. The equations of optimal foraging theory provide explicit advice to the bird, depending on whether it is operating as an energy maximizer or a time minimizer, the two objective functions utilized in many foraging models (Charnov and Orians 1973).

The theoretical behavioral ecology literature provides explicit solutions to a variety of similar problems, but it is difficult to imagine that genetic programs code for an ability to evaluate and balance so many interacting factors. Clearly, genes code for plastic rather than rigid responses, and for the use of rules of thumb, approximations that require evaluation of much less information but nonetheless yield reasonably effective responses (Stephens and Krebs 1986). However, this assumption does not provide much guidance for identifying the mechanisms by which plasticity is utilized to do "the right thing" (Emlen et al. 1998).

This problem has been recognized for about fifty years (Waddington 1957).

C. H. Waddington was concerned with both how genetic instructions could be compatible with the facts that a particular genotype might correspond to several phenotypes and that phenotypically almost identical individuals could have strikingly different genotypes. He noted that no adaptive-genetic mechanisms had been proposed to account simultaneously for the plasticity and the rigidity of canalization. The problem he posed remains a challenge (Rollo 1995; Raff 1996).

Phenotypic Plasticity

The primary attempt to resolve Waddington's dilemma, which combines plasticity with development and allometry, is based on the developmental reaction norm, the range of developmental paths a genotype can take when exposed to a variety of environments (Schlichting and Pigliucci 1998). Natural selection is presumed to act on these norms such that the organism "pulls the right levers" when confronted with a given environment. As proposed by C. D. Schlichting and M. Pigliucci, the concept involves genetic assimilation, but this is not a necessary component of the general idea. The critical requirement is plasticity integration, the matching of changes in one trait by changes in other traits. How to detect plasticity matching is as yet unresolved.

Although the developmental reaction norm hypothesis raises a variety of interesting questions, some of which are amenable to experimental testing, we are still a long way from understanding how a foraging bird is able to make appropriate decisions about which of an array of prey to eat. In short, our understanding of phenotypic evolution is weak (Orr 1999), a fact that tends to be obscured by the spectacular successes of molecular biology. Thus, one of the most important challenges faced by behavioral and evolutionary biologists is to explain the developmental processes by which the flexible but adaptive behavior of organisms actually evolves.

The Development of Human Behavior

We are, of course, especially interested in our own behavior. Analyses of human behavior are both easier and more difficult than analyses of the behavior of other species. They are easier because we can ask a richer array of questions of people and can register a richer array of answers (including verbal) than are possible with other species. They are harder because verbal responses are notoriously difficult to interpret, because of the problems in controlling for deceit and self-deception, because ethical concerns restrict the experiments possible with humans, because our behavior is unusually complicated, and because it is such a challenge to observe ourselves objectively. Religious and political baggage accompanies our attempts to understand human behavior; an evolutionary view of life challenges many deeply held beliefs (Cziko 1995; Dennett 1995).

Nonetheless, the need to better understand the evolutionary roots of human behavior is great. An evolutionary perspective is having a major impact on medicine, a field that has been remarkably nonevolutionary for most of its history (Ewald 1994; Nesse and Williams 1998). These changes include new perspectives on the evolution of pathogens, the significance and treatment of symptoms, and the design of rooms in hospitals (Ulrich 1984, 1993).

The profound influence of humans on biogeochemical cycles, environmental contamination, species extinctions, habitat fragmentation, and landscape modification gives urgency to improving our understanding of the bases of the behaviors that result in these dramatic environmental effects. Involvement with nature is strongly motivating to people in many ways. These powerful emotions influence how we respond to nature, how we attempt to manipulate it, and why we care about it. Human perceptions of and responses to environments are embedded in a complex nexus of symbolisms and cultural memories (Schama 1995). These complexities make it difficult to untangle the respective influences of our cultural and evolved responses to environments, but techniques exist that are yielding considerable progress.

One approach is based on the fact that emotional responses profoundly influence human decisions. Evolutionary biologists expect emotional responses to evolve such that they foster responses that enhance fitness. For example, those of our ancestors who did not enjoy eating and sex, and hence did not seek out food and sexual partners, presumably were more poorly represented genetically in subsequent generations than our ancestors who did enjoy those behaviors. Similar arguments apply to selection of inferior environments in which to live (Orians 1998).

Humans have both biophobic and biophilic responses to nature (Wilson 1984). It is interesting that experimental underpinnings of biophobic responses are stronger than those for biophilic responses. A major reason for the difference is that positive conditioning studies are usually more difficult to perform than aversive conditioning experiments. Environmental psychologists have carried out a series of imaginative experiments that provide interesting insights into how exposure to stimuli influences both the acquisition and the retention of fears. These experiments typically involve comparisons between aversive responses conditioned to slides of fear-relevant stimuli (snakes or spiders) and those conditioned to fear-neutral stimuli (geometric figures). Responses are typically assessed by autonomic nervous system indicators, such as skin conductance and heart rate.

A general result of these studies, most of which have been carried out in Scandinavia, is that conditioned responses to fear-relevant stimuli are always more resistant to extinction than conditioned responses to fear-neutral stimuli (McNalley 1987). To demonstrate that these differences are not the results of prior cultural reinforcement, investigators compared aversive responses to snakes and spiders with responses to far more dangerous, strongly culturally conditioned modern stimuli, such as handguns and frayed electrical wires. Conditioned aversive responses to these latter stimuli extinguished more quickly than those to snakes and spiders (Cook, Hodes, and Lang 1986; Hugdahl and Karker 1981). In addition, aversive responses to fear-relevant natural stimuli can be acquired merely by telling a person that shock will be administered. Aversive responses to fear-irrelevant natural stimuli cannot be elicited in this manner (Hugdahl 1978).

Even more striking from an evolutionary point of view are the results of "backmasking" experiments in which slides are displayed subliminally (for fifteen to thirty milliseconds) before being masked by slides of other stimuli. Even though subjects are not consciously aware of having seen the stimulus slide, presentations of slides that contain snakes or spiders elicit strong aversive reactions in normal or nonphobic persons (Öhman 1986; Öhman and Soares 1994).

These remarkable studies are just one component of investigations into varied aspects of human behavioral ecology. We have only begun to scratch the surface. Much remains to be learned about our responses to biodiversity, patterns of landscapes, and flowers (Orians 1998). We know that experiences in nature are restorative, but the mechanisms that underlie these responses are unknown (Ulrich and Addoms 1981; Kaplan and Talbot 1983; Owens 1988; Ulrich, Dimberg, and Driver 1991).

HOW DOES BEHAVIOR INFLUENCE THE STRUCTURE AND FUNCTIONING OF ECOLOGICAL SYSTEMS?

Population growth rates result from the reproductive behavior of many individuals as modified by weather, conspecifics, predators, and parasites. Behavior of individuals strongly influences the composition of ecological communities. Although it is easy to state these obvious truths, determining how individual behaviors produce their ecological consequences is a daunting task. I focus on two components of behavioral ecology whose ecological implications have been most thoroughly explored—foraging and habitat selection. These components illustrate useful approaches for integrating behavior and ecology.

Foraging and the Structure of Ecological Communities

Optimal foraging theory has for several decades been widely and successfully employed in the study of autecology (Stephens and Krebs 1986), but only recently has it been used to address ecological relationships such as competition, predation, mutualism, and trophic dynamics (Werner 1977; Belovsky 1984). Indeed, competitors, predators, and their prey, as characterized in standard competition and predator-prey equations, generally lack interesting behavior. Also, most models of metapopulation dynamics assume a constant probability of dispersal and recruitment, whereas adaptive dispersal and recruitment lead to different predictions about population dynamics (Fryxell and Lundberg 1998).

Real predators make many decisions about where to forage, which prey to pursue and capture, how much of a prey item to eat, when to leave a patch to seek alternative foraging sites, and whether to bypass a suboptimal patch to continue searching for a better one. They may form search images that increase the encounter rates with some prey types at the price of reducing encounter rates with others. Ornithologists and herpetologists differ strikingly in what they see in the same environment! In structurally complex environments they may use different search modes, concentrating on specific structural elements. Individuals of some species habitually use individuals of other species as prey finders or flushers. Some, such as obligate army-ant-following birds, are so dependent on the activities of individuals of other species, in this case ants, that they cannot survive and reproduce without them.

Biologists have barely begun to explore the ecological consequences of the behavioral complexity of real foragers in varied environments. The influence of individual behaviors on population dynamics and stability has been explored theoretically by J. M. Fryxell and P. Lundberg (1998). G. E. Belovsky (1986a, 1986b) used linear programming models of foraging by herbivores of varied sizes to assess whether herbivores regularly compete for resources, a highly contentious topic in ecology (Hairston, Smith, and Slobodkin 1960). He based his analyses on constraints on foraging behavior determined by minimum digestibility of different plant species, sizes and abundance of food items (which determine cropping rates), capacity of the animal's digestive system, and turnover rate of food in the gut. His analysis, combined with field experiments, suggests that the structure of herbivore communities may be strongly influenced by foraging behavior and foraging energetics.

A pervasive result of most empirical tests of foraging models is that, although empirical results on average support theoretical predictions, animals often do not match any single predicted optimum exactly. Instead they exhibit broad variation in performance (Ward 1992). Because the optimization models being tested predict a single optimum strategy that all individuals should adopt, the existence of substantial variation is troubling. One explanation is to interpret such variation as the simple noise about some expected value that characterizes all biological processes. However, if a high fitness premium is associated with behaving optimally, much less variation would be expected.

Another explanation is that the relevant optimality models are too simple. They focus on a single component of behavior, whereas the complex decision-making situations animals really face in nature prevent them from making truly informed choices. The proposed solution from this perspective is "satisficing"; that is, animals satisfy pressing demands as they arise (Ward 1992, 1993). For a while this idea had considerable appeal, but it has been largely abandoned because satisficing does not appear to be an evolutionarily stable strategy.

A third explanation is that animals are limited by evolved traits that prevent them from responding appropriately to the hypothesized optimization problem (Sih and Gleeson 1995). For example, there may be phylogenetic or genetic limitations that prevent animals from exhibiting appropriate antipredator behavior. This perspective suggests that it may be profitable to focus on how limiting factors render behaviors ineffective.

Recently another technique known as multiobjective programming (MOP), which has been developing for decades (see Cohon 1978 for an overview), is being employed to provide new insights into the great variability in behavior. The key idea in MOP is derivation of a nondominated or efficient set of compromises to an optimization problem that involves trade-offs among conflicting objectives. A solution is said to be nondominated or efficient if there are no other solutions that increase the contribution to one objective without resulting in a lesser contribution to at least one other objective. For example, a nondominated solution could arise when a forager reduced its nutrient intake rate to reduce its risk of predation. A dominated or inefficient decision would be one in which a forager selected a diet that did not increase its net energy intake rate but did increase its risk of predation. The MOP framework offers a quantitative way to distinguish between decision options that are or are not efficient. Multiobjective programming models allow that an array of strategies, rather than a single one, may be optimal.

For example, O. J. Schmitz and colleagues (1998) reanalyzed, from an MOP perspective, data obtained by Grantham, Morehead, and Willig (1995) on the foraging of a snail on two species of aquatic macrophytes. O. K. Grantham and colleagues used linear programming formulations to predict the optimal energy-maximizing and time-minimizing diets of snails under the conditions of their experiments. Considerable variation in foraging behavior existed among individual snails, and, contrary to expectations, most of the snails were neither energy maximizers nor time minimizers. The authors suggested that the snails exhibited partial preferences for the alternative food types, a result that is suboptimal under other foraging models. The MOP analysis by Schmitz and colleagues suggested that many diets could be efficient compromises to the trade-off between time and energy.

Insightful though this analysis was, it leaves unexplained why individual snails differed so greatly in their dietary choices. Additional experiments are needed to help us understand two kinds of variation typically observed in foraging studies. One is the deviation in average performance of individuals from some predicted optimum. The other is variability over time in individuals' performance. Theoretical analyses typically have assumed that all animals are alike and that they perceive the problems we pose to them the same way. This assumption was perhaps inevitable during the early development of foraging theory, but it is surprising that it has persisted for so long (Wilson 1998). No serious student of, say, mating behavior, would assume that all individuals are alike and treat variation in performance as uninteresting noise!

A Bayesian perspective may be helpful here (Olsson and Holmgren 1998). Individuals differ both genetically and as a result of their prior experiences in the expectations they bring to an experiment. This suggests that information about morphology and the histories of individuals should be gathered and incorporated into predictions about their varying responses to the experiments. For example, the analyses of D. S. Wilson (1998) have shown how morphological and behavioral variability within single populations of bluegill sunfish determines the variable responses of individuals to environmental contingencies. How animals perceive experimental situations, how their perceptions vary over time, and the possible genetic underpinnings of this variability remain major challenges for future research.

Habitat Selection

Habitat selection theory is closely allied to foraging theory because most investigators have used resource availability as the primary measure of habitat quality. The first explorations of habitat selection theory (Fretwell and Lucas 1969) dealt with density-dependent selection of habitats by individuals of a single species that were assumed to have perfect information upon which to base their decisions (the ideal free distribution). Subsequent theoretical elaborations dealt with the costs of acquiring information, and the amount of time an individual gets to use different patches. An interesting result of these analyses is that anything that promotes a difference in reward structure among patch types favors habitat selection (Rosenzweig 1986). Methods have been developed to assess density-dependent habitat selection in species that are difficult to observe directly (Rosenzweig and Abramsky 1984).

Using habitat selection theory, L. R. Lawlor and J. Maynard Smith (1976) explored relationships between competition and habitat selection. To achieve mathematical tractability, they limited their analysis to a two-patch, two-species system in which the species differed over which patch they preferentially used. The theoretical result concluded that under conditions of competition each species became a specialist on one habitat type. A powerful graphical method for approaching habitat selection under competition uses a technique known as isoleg analysis, after the Greek words for equal (*isos*) and choice (*lego*). An isoleg is a line in a state space of animal densities such that some aspect of a species' habitat selection is constant at every point on the line (Rosenzweig 1981).

The first isoleg analyses confirmed the results of Lawlor and Maynard Smith (Rosenzweig 1981). Subsequent analyses, which added interference competition, yielded the surprising result that individuals of a subordinate species may actually preferentially select a less optimal habitat in the presence of a competitor (Pimm and Rosenzweig 1981). A field test of habitat selection among hummingbirds confirmed these predictions (Pimm, Rosenzweig, and Mitchell 1985). The general predictions from isoleg theory are all qualitative, but, given the complexities of ecological interactions, understanding qualitative patterns in the structure of ecological communities is a worthy goal.

A major challenge for future research is to determine which types of adaptive behaviors have the greatest influence on population dynamics and on the structure and functioning of ecological communities. The richness of potential outcomes is signaled by the theoretical analyses of J. M. Fryxell and P. Lundberg (1998), which suggest that age- or size-dependent feeding selectivity probably has little effect on population dynamics, whereas almost all forms of interference and territoriality are strongly stabilizing. We need similar analyses that deal with ecological features in addition to population dynamics.

DOES BEHAVIOR DRIVE OR CONSTRAIN EVOLUTION?

Various views have been advanced concerning the influence of behavior on rates and direction of evolution. On the one hand, Ernst Mayr (1974) pointed out that different kinds of behaviors play different roles in evolution. He also suggested that new selection pressures, induced by changes in behavior, might lead to morphological changes that facilitate the occupation of new ecological niches. In his view behavior probably played an important role in stimulating major evolutionary radiations.

On the other hand, behavioral conservatism could prevent organisms from responding appropriately to environmental changes. This would be true if some of the ghosts of selection past persist for a long time. Considerable evidence suggests that organisms' perceptual systems have highly conserved functional properties, such as the ability to recognize and respond to ecologically important habitat features and the traits of predators that were formerly, but no longer are, present in the environment (Coss 1991; Curio 1993). Current mate-selection behavior may be influenced by responses of individuals in past generations to potential mates (Ryan and Rand 1993). Thus, pattern recognition responses may persist for thousands of generations if natural selection on the traits is weak. Nevertheless, our understanding of the resistance of behavioral traits to extinction when they are no longer favored or when selection against them is weak is highly fragmentary. Despite these difficulties modern methods enable investigators to examine the potential role of behavior on rates and modes of speciation and whether behavior might influence rates of speciation.

Whether speciation in most lineages, with the obvious exception of polyploidy, requires geographic isolation, has long been an issue of contention in evolutionary biology. The major obstacle to sympatric speciation appears to be the homogenizing effect of recombination. By regularly bringing together genes from two parts of a population that are adapting to different habitats, recombination is believed to prevent natural selection from favoring genes and gene combinations that would otherwise be favored in the different habitats. Behavior has figured prominently in attempts to find ways to offset the homogenizing influence of recombination. Searchers for its mechanisms have proposed that behavior could permit sympatric speciation by means of both habitat selection and mate selection.

Theoretical analyses suggest that sexual selection, which is a powerful, diversifying process, may promote rapid speciation between disjunct populations and those in secondary contact (Lande 1982; West-Eberhard 1983). This theoretical result is supported by the fact that in birds, species richness is greater in polygynous than in monogamous lineages judged to be of the same age (Mitra, Landel, and Pruett-Jones 1996), but many more phylogenetic studies of this type are needed to assess the generality of the pattern.

Theoretical analyses also suggest that strong habitat fidelity can favor sympatric speciation under reasonable intensities of selection and penetrance for habitat selection and nonhabitat assortative mating. For example, a model developed by P. A. Johnson and colleagues (1996) suggests that habitat preference genes and habitat-specific fitness genes can become associated when assortative mating occurs because of habitat preference, provided that habitat preference is nearly error-free. However, their model suggests that, if coupled with nonhabitat assortative mating, habitat preferences can eliminate interbreeding. Whether or not sexual selection results in high rates of speciation may depend on which sex exerts the most powerful selective role. Males, by moving more and courting females relatively indiscriminately, tend to favor gene flow. The more sedentary, discriminating females generate barriers to gene flow (Parker and Partridge 1998).

Many laboratory experiments with *Drosophila* have subjected populations to disruptive selection designed to test habitat isolation models of sympatric speciation (Rice and Hostert 1993). These experiments show that sympatric speciation is possible in principle, but it remains to be determined whether sufficiently intense disruptive selection regularly exists in nature. Thus, the resolution of one of the most durable controversies in evolutionary biology will require imaginative behavioral experiments in the field.

A number of investigators have suggested that behavioral evolution should influence extinction rates, but they have not agreed on how. One possibility is that sexual selection could reduce extinction rates because, by choosing mates with greater stamina, females may reduce the risk of acquiring sexually transmitted diseases (Hamilton and Zuk 1992), may gain better genes for their offspring, and may garner more assistance from their mates. Alternatively, sexual selection might result in reduced population fitness because it constrains phenotypic plasticity or responses to natural selection pressures (Lande 1980; Kirkpatrick 1996). Empirical tests of these conjectures are difficult to devise, but introductions of species to oceanic islands offers one fruitful source of information. D. K. McLain, M. P. Moulton, and J. G. Sanderson (1999) compared the survival of plumage-dimorphic and plumage-monomorphic species of birds introduced to oceanic islands. The establishment success rate was less for species with sexually dichromatic plumage. These results are consistent with the view that sexual selection indirectly promotes extinction of small, colonizing populations that encounter new environmental conditions.

DETERMINING THE COURSE OF EVENTS LEADING TO PAST AND CURRENT BEHAVIOR PATTERNS

In addition to studying the adaptive significance of behavior and its implications for ecology, biologists are interested in the historical pathways by which current behavioral repertories evolved. Progress in this field has been slow because behavior leaves few traces in the fossil record. Fossil evidence has been used to determine when organisms first evolved the ability to burrow and to swim above the substratum. Running style and speed have also been inferred from fossil footprints. Fossil evidence strongly suggests that some dinosaurs performed elaborate parental care. It is interesting that sexual dimorphism in size and feather traits has indicated that a lek mating system probably characterized *Confuciusornis sanctus*, one of the earliest birds (Martin et al. 1998).

These successes notwithstanding, the tool that has catalyzed the most striking progress in our understanding of the evolution of behavior has been the development of powerful methods of inferring phylogenies (Edwards and Naeem 1993; Winkler and Sheldon 1993; Edwards and Naeem 1994; Gess 1996; Lee et al. 1996). Given a plausible phylogeny, investigators can infer the probable ancestral state of traits, the number of times different behavioral traits evolved, the lability and speed of evolutionary changes in behavior, and constraints on behavioral evolution.

There are, of course, pitfalls associated with the use of cladistic methods. All phylogenies are provisional hypotheses about evolutionary relationships that may be revised as new evidence becomes available. Often substantially different phylogenies are approximately equally parsimonious based on existing data. Nonetheless, provisional phylogenies are useful because, in addition to their use to interpret the history of traits, sensitivity analyses can investigate how alternative phylogenies would affect interpretation of trait histories.

A second problem associated with using phylogenies to infer patterns of evolution is to determine which of several possible methods is most appropriate. One approach is to identify a trait shared by members of a species-rich clade and then make a plausibility argument for its effect on the rates of cladogenesis in the lineage. A second approach is to compare the diversity and trait values of several clades of the same rank. A third method is to compare sister clades that differ in diversity and in the trait being analyzed. The answer one gets may depend on the method employed. For example, when A. O. Mooers and A. P. Moller (1996) used a tribal level of analysis, they found a significant correlation between coloniality and rate of speciation. However, they rejected the tribal level of analysis on the ground that sample size would be inflated if tribes with both many species and a high proportion of colonial species were close relatives. Therefore, they chose a sister group comparison to test the hypothesis that birds which breed colonially might be under stronger sexual selection and have faster rates of evolution, and might, therefore, speciate more rapidly than those that do not breed colonially. This analysis found no association between the number of species in a clade and whether or not it was colonial.

Michael Rosenzweig (1996), however, noted that the sister group method reduced the number of data points available for analysis from 175 (if a tribal analysis were employed) to 25 or 13 for the two sister group tests used by Mooers and Moller. Given that the potential relationship between coloniality and speciation needed to be distinguished from the potentially confounding role of body size, the greatly reduced statistical power of the sister group method substantially increased the probability of a Type II error.

In essence, the sister group method assumes that the only useful information is found at the nodes of a cladogram. However, given that species can, and do, evolve in both directions between coloniality and noncoloniality, the descendants of colonial species that remain colonial may indicate something important about evolution. Given that an analysis comparing tribes did suggest a relationship between coloniality and speciation rates, the issue remains unresolved.

Because of the relatively greater probability of reversibility in behavioral than in morphological traits, interpretational problems are more complex for analyses of behavioral than for morphological traits. Consider, for example, the evolution of brood parasitism in cowbirds (Lanyon 1992). The five species of parasitic cowbirds exhibit a variety of host specificity, ranging from a species that parasitizes only one host to a species with more than two hundred known hosts. An analysis of DNA sequences of the mitochondrial cytochrome-*b* genes confirmed that the brood-parasitic cowbirds represent a monophyletic assemblage. An exhaustive search identified a single most parsimonious tree. The deeper splits in the phylogeny separate species that parasitize few species of hosts. The two species that parasitize many hosts form a recently split sister species. Based on this information, Lanyon (1992) speculated that the generalized form of brood parasitism is a derived condition that evolved from host-specific brood parasitism. This interpretation is plausible, but it is not the only interpretation consistent with the phylogeny. In fact, the reverse, that the ancestral state was generalized parasitism, is also consistent with the data. This alternative interpretation views the phylogeny as showing that the longer a lineage has been evolving as a brood parasite, the fewer species the members of that lineage parasitize. Other evidence supports Lanyon's interpretation, but the issue is unresolved. Thus, even with an agreed-upon phylogeny, inferring patterns in the evolution of behavior is difficult.

W. A. Searcy, K. Yasukawa, and S. Lanyon (1999) used a phylogeny of a lineage of American blackbirds to infer the pattern of evolution of polygyny. They mapped behavioral traits on a phylogeny, which was based on DNA data, and assumed that the traits were unordered, that is, a change between any two states of a character was possible in a single step. The traits they used were mating relationships (monogamous, polygynous, or promiscuous), spacing behavior (territorial or nonterritorial), nesting dispersion (dispersed or colonial), nesting habitat (marsh, grassland, or woodland), and parental care (equal by both sexes, unequal, female only, or none). Their analysis suggested how many times polygyny evolved within the blackbird lineage and that polygyny evolved from monogamy. The data also suggested that territoriality, nesting in woodland habitat, and equal parental care by both sexes are ancestral states within the lineage. These conclusions have implications for a wide array of hypotheses about the evolution of social behavior in this and other lineages of birds as well.

One approach to inferring ancestral states of behavioral traits is to use concepts of directionality that may suggest constraints on evolution. Evolutionary irreversibilities, or reversals that are extremely difficult to achieve, can arise from several causes. One is that intermediates may be more difficult to enter from one direction than another. For example, complex behaviors, structures, and physiological systems probably are more easily lost than gained. It is probably easier to lose limbs than to regain them. Snakes are probably committed to a limbless life as long as serpents survive. Once a brood parasite has lost the full array of parental behaviors, it may be impossible or nearly impossible to reevolve them. A phylogenetic analysis of avian parental care indeed suggests that biparental care is the ancestral state among birds (McKitrick 1992). Nevertheless, incubation of eggs did evolve from ancestors that did not incubate; that is, complex behaviors can and do evolve.

Second, a state may bring about secondary changes that make reversal difficult. Thus, the evolution of sex chromosomes led to the degeneration of the Y chromosome. Regenerating a Y chromosome with a full complement of genes may be impossible. Similarly, a state may become absorbing. For example, the evolution of sex does not prevent the splitting off of asexual lineages, but evidence suggests that they do not persist very long. Becoming dioecious may prevent a plant lineage from reevolving hermaphroditism. A major challenge in the study of evolution is to identify the importance of such constraints on the direction of behavioral evolution.

The ontogeny of behavior can also be used to help infer ancestral states. Especially interesting are products of behavior, such as nests, as extended phenotypes. Nests are especially informative because, in addition to the nest, the sequence of behavioral acts to construct it may reveal further information about the homology of cryptically homologous or superficially similar but analogous nests. Ontogenetic criteria can be used to determine the polarity of nest characters because of the physical or engineering constraints on the development of three-dimensional structures. For example, no avian nests are constructed first and subsequently attached to a substratum.

       A   B

Figure 5.1. Nests of ovenbirds (Furnariidae). **A.** A leñatero (*Anumbius annumbi*) nest in Argentina. **B.** A nest of *Pseudoseisura lophotes* in a *Trichocereus* cactus in Argentina.

    K. Zyskowski and R. O. Prum (1999) used the tremendous architectural diversity of nests of neotropical ovenbirds (family Furnariidae), combined with the ontogeny of nest construction, to derive a phylogeny of the ovenbirds (Figures 5.1 and 5.2). Their analysis argued for some revisions in the generally accepted phylogeny of the group, resolved some ambiguities, and suggested profitable avenues for future research, which may, of course, require revision of their tentative phylogenetic reconstruction. Thus, phylogenetic analyses can be used to infer patterns of behavioral evolution, and behavior can, in turn, be used in the construction of phylogenies. How conflicts between phylogenies suggested by behavior and those suggested by molecular data should be resolved remains unclear, although most investigators place greater trust in molecular data.

CONCLUDING REMARKS

Identifying the processes that lead to the unfolding of a functioning adult from a single-celled zygote is a major challenge for modern biology. How organisms are able to do the right thing is perhaps the most complex component of developmental processes, both because behavior is intrinsically rich and varied and

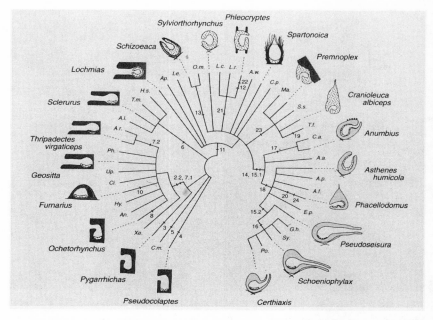

Figure 5.2. Phylogeny of ovenbirds (Furnariidae) based on nest structure (from Zyskowski and Prum 1999).

because of the organisms' extensive learning capabilities. We are still relatively ignorant of how genes actually influence behavior in general and learning in particular. Until recently it was believed that neurons in the central nervous system could be lost but not regained. However, investigations of song learning in birds (Nottebohn 1985) have shown that new neurons can form in some but not all parts of the avian central nervous system. Cell death may function as a trigger for the generation of new neurons, but as yet we have only slight hints about the process.

Ignorance of the role of behavior in the structure and functioning of ecological communities is equally vast. Sixteen years ago Michael Rosenzweig (1985) posed a set of questions about community structure. Among them were: Why is exploitative competition apparently rare among habitat selectors? Under what conditions should we expect competitive interactions to be asymmetrical (as they often are)? Why should one congener be social whereas another in the same or similar habitat is often asocial? How does predation influence habitat selection? Our ability to answer these questions, all of which concern how organisms behave, has not improved much since Rosenzweig posed them.

The general answer to the question Does behavior drive or constrain evolution? is, of course, that it does both. The challenge is to determine which components of evolution have been stimulated by behavior and for which ones behavior has been a serious constraint. The central problem is to identify which types of behavior are most resistant to change and which are difficult to reverse. Recent developments in inferring phylogenies are providing the first major opportunities to approach these problems. Nevertheless, extreme caution is advised, because all phylogenies are provisional. Most of them are based on a limited number of traits or a tiny fraction of the total genome. Typically a number of phylogenies are nearly equally parsimonious. As Jonathan Coddington (1992) stressed, if a phylogeny is wrong, analyzing how it came about is pointless. Unfortunately, a considerable amount of confusion is likely to be generated by overzealous interpretations of the implications of provisional phylogenies.

Finally, we urgently need to incorporate human biology fully into an evolutionary framework. Better understanding of the evolutionary roots of our behavior not only is of great theoretical interest but has important policy applications, such as the design of hospital rooms (Ulrich 1993), buildings in general, and counseling for stepparents (Daly and Wilson 1985, 1996). With improved understanding we may be able to comprehend some puzzling human propensities. For example, why do we have a passion for collecting things? Why are we uncomfortable with windowless rooms? Why are we so fond of flowers? Why do we fall in love?

And why do we attach great significance to purely arbitrary moments in time? On my seventieth birthday, which will come sooner than I would wish, I will be, as I have been every other day of my life, one day older, but I am unlikely to see it that way. I will reflect on the passing of another decade in my life, as if a decade had some intrinsic meaning. And what elements of our behavioral constitution have brought us together at the millennium's end to celebrate an arbitrary moment in time whose reference base is almost certainly incorrect?

ACKNOWLEDGMENTS

The clarity and organization of the manuscript were greatly improved by comments from Ray Huey, Sievert Rohwer, and two anonymous reviewers, none of whom is, of course, responsible for its general content.

LITERATURE CITED

Belovsky, G. E. 1984. Snowshoe hare optimal foraging and its implications for population dynamics. *Theoret. Popul. Biol.* 25:235–64.

———. 1986a. Optimal foraging and community structure: Implications for a guild of generalist grassland herbivores. *Oecologia* 70:35–52.

———. 1986b. Generalist herbivore foraging and its role in competitive interactions. *Amer. Zool.* 26:51–69.

Brower, L. P. 1977. Monarch migration. *Nat. Hist.* 86:40–53.

Charnov, E. L., and G. H. Orians. 1973. *Optimal foraging: Some theoretical explorations.* Dept. of Biology, Univ. of Utah.

Coddington, J. A. 1992. The comparative method in evolutionary biology. *Trends Ecol. Evol.* 7:68–69.

Cohon, J. L. 1978. *Multiobjective programming and planning.* New York: Academic Press.

Cook, E. W., R. L. Hodes, and P. J. Lang. 1986. Preparedness and phobia: Effects of stimulus content on human visceral conditioning. *J. Abnormal Psych.* 95:195–207.

Coss, R. G. 1991. Evolutionary persistence of memory-like processes. *Concepts in Neuroscience* 2:129–68.

Curio, E. 1993. Proximate and developmental aspects of antipredator behavior. *Adv. Study of Behav.* 22:135–237.

Cziko, G. 1995. *Without miracles: Universal selection theory and the second Darwinian revolution.* Cambridge, Mass.: MIT Press.

Daly, M., and M. I. Wilson. 1985. Child abuse and other risks of not living with both parents. *Ethol. & Sociobiol.* 6:155–76.

———. 1996. Violence against stepchildren. *Current Directions in Psychol. Science* 8:155–76.

Dennett, D. C. 1995. *Darwin's dangerous idea.* New York: Simon & Schuster.

Edwards, S. V., and S. Naeem. 1993. The phylogenetic component of cooperative breeding in perching birds. *Amer. Nat.* 141:754–89.

———. 1994. Homology and comparative methods in the study of avian cooperative breeding. *Amer. Nat.* 143:723–33.

Emlen, J. M., D. C. Freeman, A. Mills, and J. H. Graham. 1998. How organisms do the right thing: The attractor hypothesis. *Chaos* 8:717–26.

Ewald, P. W. 1994. *Evolution of infectious disease.* New York: Oxford Univ. Press.

Fretwell, S. D., and H. L. Lucas, Jr. 1969. On territorial behavior and other factors influencing habitat distribution in birds: I. Theoretical development. *Acta Biotheoretica* 19:16–36.

Fryxell, J. M., and P. Lundberg. 1998. *Individual behavior and community dynamics.* New York: Chapman & Hall.

García, J., and R. A. Koelling 1966. Relation of cue to consequence in avoidance learning. *Psychonomic Science* 4:123–24.

Gess, S. K. 1996. *The pollen wasps: Ecology and natural history of the Masarinae.* Cambridge, Mass.: Harvard Univ. Press.

Goldsmith, T. H. 1990. Optimization, constraint, and history in the evolution of eyes. *Q. Rev. Biol.* 65:281–322.

Grantham, O. K., D. L. Morehead, and M. R. Willig. 1995. Foraging strategy of the giant ramshorn snail, *Marisa cornuarietis*: An interpretive model. *Oikos* 72:333–42.

Hairston, N. G., F. E. Smith, and L. B. Slobodkin. 1960. Community structure, population control, and competition. *Amer. Nat.* 94:421–25.

Hamilton, W. D., and M. Zuk. 1992. Heritable fitness and bright birds: A role for parasites. *Science* 218:384–87.

Hugdahl, K. 1978. Electrodermal conditioning to potentially phobic stimuli: Effects of instructed extinction. *Behav. Res. & Therapy* 16:315–21.

Hugdahl, K., and A. Karker. 1981. Biological vs. experiential factors in phobic conditioning. *Behav. Res. & Therapy* 18:109–15.

Hutchinson, G. E. 1965. *The ecological theatre and the evolutionary play*. New Haven: Yale Univ. Press.

Johnson, P. A., F. C. Hoppensteadt, J. J. Smith, and G. L. Bush. 1996. Conditions for sympatric speciation: A diploid model incorporating habitat fidelity and non-habitat assortative mating. *Evol. Ecol.* 10:187–205.

Kaplan, S., and J. F. Talbot. 1983. Psychological benefits of a wilderness experience. In *Behavior in the natural environment*, edited by I. Altman and J. F. Wohlwill. New York: Plenum Press.

Kirkpatrick, M. 1996. Good genes and direct selection in the evolution of mating preferences. *Evolution* 50:2125–40.

Lande, R. 1980. Sexual dimorphism, sexual selection, and adaptation in polygenic characters. *Evolution* 34:292–307.

———. 1982. Rapid origin of sexual isolation and character divergence in a cline. *Evolution* 36:213–23.

Lanyon, S. M. 1992. Interspecific brood parasitism in blackbirds (Icterinae): A phylogenetic perspective. *Science* 255:77–79.

Lawlor, L. R., and J. Maynard Smith. 1976. The coevolution and stability of competing species. *Amer. Nat.* 110:79–99.

Lee, P. M. L., D. H. Clayton, R. Griffiths, and R. D. M. Page. 1996. Does behavior reflect phylogeny in swiftlets (Aves: Apodidae): A test using cytochrome *b* mitochondrial DNA sequences. *Proc. Natl. Acad. Sci.* 93:7091–96.

Lima, S. L. 1998. Nonlethal effects in the ecology of predator-prey interactions. *BioScience* 48:25–34.

McKitrick, M. C. 1992. Phylogenetic analysis of avian parental care. *Auk* 109:828–46.

McLain, D. K., M. P. Moulton, and J. G. Sanderson. 1999. Sexual selection and extinction: The fate of plumage-dimorphic and plumage-monomorphic birds introduced onto islands. *Evol. Ecol. Res.* 1:549–65.

McNalley, R. J. 1987. Preparedness and phobias: A review. *Psychol. Bull.* 101:283–303.

Mangel, M., and C. W. Clark. 1988. *Dynamic modeling in behavioral ecology*. Princeton, N.J.: Princeton Univ. Press.

Marchetti, C. 1998. Notes on the limits of knowledge explored with Darwinian logic. *Complexity* 3:22–35.

Martin, L. D., Z. Zhou, L. Hou, and A. Feduccia. 1998. *Confuciusornis sanctus* compared to *Archaeopteryx lithographica*. *Naturwissenschaften* 85:286–89.

Maynard Smith, J. 1978. Optimization theory in evolution. *Ann. Rev. Ecol. Syst.* 9:31–56.

Mayr, E. 1960. The emergence of evolutionary novelties. *Taxa* 1960:349–80.

———. 1974. Behavior programs and evolutionary strategies. *Amer. Sci.* 62:650–59.

Mitra, S., H. Landel, and S. Pruett-Jones. 1996. Species richness covaries with mating system in birds. *Auk* 113:544–51.

Mooers, A. O., and A. P. Moller. 1996. Colonial breeding and speciation in birds. *Evol. Ecol.* 10:375–85.

Nesse, R. M., and G. C. Williams. 1998. Evolution and the origins of disease. *Sci. Amer.,* Nov., 32–39.

Nottebohn, F. 1985. Neuronal replacement in adulthood. *Ann. N.Y. Sci.* 457:143–59.

Öhman, A. 1986. Face the beast and fear the face: Animal and social fears as prototypes for evolutionary analyses of emotion. *Psychophysiology* 21:123–45.

Öhman, A., and J. J. F. Soares. 1993. Unconscious anxiety: Phobic responses to masked stimuli. *J. Abnormal Psychol.* 103:231–40.

Olsson, O., and N. A. Holmgren. 1998. The survival-rate-maximizing policy for Bayesian foragers: Wait for good news. *Behav. Ecol.* 9:345–53.

Orians, G. H. 1998. Human behavioral ecology: 140 Years without Darwin is too long. *Bull. Ecol. Soc. Amer.* 79:15–28.

Orr, H. A. 1999. An evolutionary dead end? *Science* 285:343–44.

Owens, P. E. 1988. Natural landscapes, gathering places, and prospect refuges: Characteristics of outdoor places valued by teens. *Children's Environmental Q.* 3:17–24.

Parker, G. A., and L. Partridge. 1998. Sexual conflict and speciation. *Philos. Trans. R. Soc. Lond. B* 353:261–74.

Penn, D. J., and W. K. Potts. 1999. The evolution of mating preferences and major histocompatibility complex genes. *Amer. Nat.* 153:145–64.

Pimm, S. L., and M. L. Rosenzweig. 1981. Competitors and habitat use. *Oikos* 37:1–6.

Pimm, S. L., M. L. Rosenzweig, and W. Mitchell. 1985. Competition and food selection: Field tests of a theory. *Ecology* 66:798–807.

Plotkin, H. C. 1988. Learning and evolution. In *The role of behavior in evolution,* edited by H. C. Plotkin. Cambridge, Mass.: MIT Press.

Potts, W. K., C. J. Manning, and E. K. Wakeland. 1991. Mating patterns in seminatural populations of mice: Influence by MHC genes. *Nature* 352:619–21.

Pulliam, H. R., and C. Dunford. 1980. *Programmed to learn: An essay on the evolution of culture.* New York: Columbia Univ. Press.

Raff, R. A. 1996. *The shape of life: Genes, development, and the evolution of animal form.* Chicago: Univ. of Chicago Press.

Rice, W. R., and E. E. Hostert. 1993. Laboratory experiments on speciation: What have we learned in forty years? *Evolution* 47:1637–53.

Rollo, C. D. 1995. *Phenotypes: Their epigenetics, ecology, and evolution.* New York: Chapman & Hall.

Rosenzweig, M. L. 1981. A theory of habitat selection. *Ecology* 62:327–55.

——. 1985. Some theoretical aspects of habitat selection. In *Habitat selection,* edited by M. L. Cody. New York: Academic Press.

——. 1996. Colonial birds probably do speciate faster. *Evol. Ecol.* 10:681–83.

Rosenzweig, M. L., and Z. Abramsky. 1984. Detecting density-dependent habitat selection. *Amer. Nat.* 126:405–17.

Ryan, M. J., and A. S. Rand. 1993. Sexual selection and signal evolution: The ghost of biases past. *Philos. Trans. R. Soc. Lond. B* 340:187–95.

Schama, S. 1995. *Landscape and memory.* New York: Alfred A. Knopf.

Schlichting, C. D., and M. Pigliucci. 1998. *Phenotypic evolution: A reaction norm perspective.* Sunderland, Mass.: Sinauer Associates.

Schmitz, O. J., J. L. Cohon, K. D. Rothley, and A. P. Beckerman. 1998. Reconciling variability and optimal behavior using multiple criteria optimization models. *Evol. Ecol.* 12:73–94.

Searcy, W. A., K. Yasukawa, and S. Lanyon. 1999. Evolution of polygyny in the ancestors of red-winged blackbirds. *Auk* 116:5–19.

Seligman, M. E. P. 1970. On the generality of the laws of learning. *Psychol. Rev.* 77:406–18.

Sih, A., and S. K. Gleeson. 1995. A limits-oriented approach to evolutionary ecology. *Trends Ecol. Evol.* 10:378–81.

Stephens, D. W., and J. R. Krebs. 1986. *Foraging theory.* Princeton, N.J.: Princeton Univ. Press.

Sutherland, W. A., and G. A. Parker. 1992. The relationship between continuous input and interference models of ideal free distribution with unequal competitors. *Anim. Behav.* 44:345–55.

Tinbergen, N. 1951. *The study of instinct.* New York: Oxford Univ. Press.

Ulrich, R. S. 1984. View through a window may influence recovery from surgery. *Science* 224:420–21.

——. 1993. Biophilia, biophobia, and natural landscapes. In *The biophilia hypothesis,* edited by S. R. Kellert and E. O. Wilson. Washington, D.C.: Island Press.

Ulrich, R. S., and D. Addoms. 1981. Psychological and recreational benefits of a neighborhood park. *J. Leisure Res.* 13:43–65.

Ulrich, R. S., O. Dimberg, and B. L. Driver. 1991. Psychological indicators of leisure benefits. In *Benefits of leisure,* edited by B. L. Driver, P. J. Brown, and G. L. Peterson. State College, Pa.: Ventura.

Waddington, C. H. 1957. *The strategy of the genes.* London: Allen & Unwin.

Ward, D. 1992. The role of satisficing in foraging theory. *Oikos* 63:312–17.

——. 1993. Foraging theory, like all fields of science, needs multiple hypotheses. *Oikos* 67: 376–78.

Werner, E. E. 1977. Species packing and niche complementarity in three sunfishes. *Amer. Nat.* 111:553–78.

West-Eberhard, M. J. 1983. Sexual selection, social competition, and speciation. *Q. Rev. Biol.* 58:155–83.

Whitehead, A. N. 1911. *Introduction to mathematics*. New York: Henry Holt.

Wilson, D. S. 1998. Adaptive individual differences within single species populations. *Phil. Trans. R. Soc. Lond. B* 353:199–205.

Wilson, E. O. 1984. *Biophilia*. Cambridge, Mass.: Harvard Univ. Press.

Winkler, D. W., and F. H. Sheldon. 1993. Evolution of nest construction in swallows (Hirundinidae): A molecular phylogenetic perspective. *Proc. Natl. Acad. Sci.* 90:5705–7.

Zyskowski, K., and R. O. Prum. 1999. Phylogenetic analysis of the nest architecture of neotropical ovenbirds (Furnariidae). *Auk* 116:891–911.

6

Conserving Biodiversity into the New Century

GHILLEAN T. PRANCE

There cannot be a more urgent topic for biologists to address in the new century than the conservation of biodiversity. Species continue to be lost at an alarming rate that is predicted to rise (see, for example, Wilson 1992; Ehrlich 1994; and Myers 1996; and Figure 6.1). Only just over 7 percent of Brazil's Atlantic forest remains in an area that is full of endemic plants and animals (Mori, Boom, and Prance 1981). In Madagascar and the Mascarenes, where most of the plants and animals are endemic, only 10 percent of the original vegetation remains. The loss of biodiversity has been well documented elsewhere, and the hot spots of biodiversity, where species are concentrated and often highly threatened, have been well defined (Myers 1990). The concept of hot spots has recently been updated, and it has been shown that twenty-five hot spots constituting only 1.4 percent of the land surface of the Earth contain as many as 44 percent of all species of vascular plants and 35 percent of all species in four vertebrate groups (Myers et al. 2000). It is not my goal here to further document the serious loss of species and habitat confronting us today but rather to discuss some of the aspects, mainly from the point of view of a systematist, that we need to continue to address if we are to save biodiversity from total destruction.

We now have the Convention on Biological Diversity signed and ratified by 172 political entities (unfortunately excluding the United States). Many biologists have been most active in the field of conservation, as witnessed by the many conservation journals founded during the last two decades of the last century, for example, *Conservation Biology, Biodiversity and Conservation,* and *Conservation Ecology,* among others. There are also a plethora of government and nongovernmental organizations and societies whose remit is conservation,

Figure 6.1. Much of the species-diverse forest of Rondônia State, Brazil, has been felled for pasture and agriculture. This center of biological diversity is fast being destroyed.

and large foundations, such as the John D. and Catherine T. MacArthur Foundation and the W. Alton Jones Foundation, have invested large funds in conservation. In spite of all this effort the destruction of biodiversity continues at a most alarming rate, so the challenge before us is enormous.

We have not yet managed to convince the world that the loss of species is a threat to the existence of our own species. The first decades of this new century probably offer the last opportunities both to understand the complexity of the biological systems that hold together life on our planet and to conserve the biological diversity and web of interactions that make up these systems. It is quite obvious that the biological community needs to be even more active and united in matters of conservation or we will not even have the biological species to study. Instead of arguing over things such as which method of cladistics to use, we need to be convincing politicians, economists, planners, and industrialists of the importance of all biological species.

THE INVENTORY IS FAR FROM COMPLETE

One of the problems facing conservationists is that the inventory of biodiversity is far from complete, especially in tropical regions. This problem has been well documented elsewhere (Prance et al. 2000), so a few botanical examples will illustrate this point here. The Index Kewensis, which has recently gone live on the Internet as part of the International Plant Names Index (IPNI, see http:/www.ipni.org), lists the names of vascular plant species described by plant taxonomists. In the nine years between 1989 and 1997, botanists of the world described 21,097 new species of vascular plants. That is, well over 2,000 new species were added each year. This is easily seen from some of the areas where the Royal Botanic Gardens, Kew, is conducting fieldwork. New species are frequently found in places such as the Atlantic forests of Brazil, Madagascar, and New Guinea. It is often the so-called highly threatened hot spots that are yielding the most new species.

Areas that appear to be well known, such as the vicinity of Manaus, Brazil (Nelson et al. 1990), are actually far from well collected. We recently completed a field guide to the Reserva Florestal Adolpho Ducke, just on the edge of the city of Manaus (Ribeiro et al. 1999). In 1993 we set out with a preliminary list of plant species from the herbarium database of the Instituto Nacional de Pesquisas da Amazonia (INPA) containing 825 species (Prance 1990). Five years later, as the project neared completion, the list was expanded to 2,175 species, including at least 50 new species (see Prance et al. 2000). This work showed the value of the intensive study of a small area of tropical forest but also how incomplete is the inventory of even relatively well collected groups of organisms. If it had been a study of the insects or fungi, many more new taxa would have been discovered. When I carried out a study of the pollination of one of the best known Amazonian plants, *Victoria amazonica*, the royal water lily (Figure 6.2), its beetle pollinator turned out to be an undescribed species of *Cyclocephala*. The age of biological exploration is far from over, if only natural ecosystems survive to be inventoried further.

The incomplete inventory is not just of species but also of vegetation types and of the complexity of interactions that hold ecosystems together. There is a certain danger of placing too much emphasis on species alone in conservation planning and in fieldwork. Knowledge of the dynamics, mutualisms, and dependencies within an ecosystem is essential for successful conservation. The study of long-term plots is yielding an abundance of useful data; this aspect needs to be encouraged and increased, especially in areas of tropical forests of all types.

CONSERVING THE ENVIRONMENTAL SERVICES OF BIODIVERSITY

While the conservation of the areas defined as hot spots will preserve a large number of species, and is well worth the large investment of funds suggested

Figure 6.2. *Victoria amazonica*, the royal water lily, whose beetle pollinator, studied in 1975, turned out to be an undescribed species of *Cyclocephala*.

by N. Myers and colleagues (2000), conservation must not be focused solely on the small area of land represented by these hot spots. As the facts about the reality of climate change become more apparent, and the world suffers more freak storms, floods, and other site anomalies, the environmental service role of natural vegetation becomes more obvious (Costanza et al. 1997; Pimm 1997). Conservation of the vegetation in hot spots will do little to help in these cases. A good example is seen in the destruction of mangrove forests. The typhoon that hit the Indian state of Orissa in October 1999 would have been much less damaging if the mangrove forests had been there to reduce its effect. The damage caused by recent floods in Venezuela or the hurricane that hit Honduras in 1998 would have been far less if the natural vegetation had remained intact. Perhaps we are placing too much emphasis on total biodiversity and not enough on the real role that natural ecosystems play in the maintenance of the balance of life on Earth. The value of large stretches of less species-diverse forests as carbon sinks and as means of climate control should also be a focus of biodiversity conservation. Without stable climate species may not survive in the hot spots even if the right measures are taken to preserve them intact. Climate change may be controlled by maintaining vast areas of forest that is far less species diverse than the hot spots.

PREDICTIVE CLASSIFICATIONS AND MOLECULAR GENETICS

Molecular techniques furnish many new data useful for the conservation of bio-diversity, and they will become increasingly important as the dilemma of what to conserve heightens. The two principal uses at present are to improve evolutionary classifications and to determine relationships within small populations.

The use of molecular systematics in establishing the evolutionary relationships between organisms is much more accurate than morphology alone. Figure 6.3 is a cladogram of the genera of the Lecythidaceae (which now includes genera formerly placed in Scytopetalaceae). The relationships expressed here are likely to be far more in line with evolution because the cladogram is based on combined molecular and morphological data. If decisions are needed about what locations should take priority to preserve the diversity of the Lecythidaceae, one would want as broad a spectrum over the cladogram as possible.

Data from the Lecythidaceae and Chrysobalanaceae have been placed into the Worldmap Program (Vane-Wright, Humphries, and Williams 1991; Williams, Vane-Wright, and Humphries 1992; Williams et al. 1996). This program includes two algorithms for the selection of priority areas for conservation in the units of area selected (e.g., degree squares), one on species of alpha diversity and the other on clusters of endemics. However, in addition to these criteria, the program displays the cladogram of the group that is the source of data. When two areas have a similar diversity or number of endemics, the information from the cladogram proves most useful. The area that has the broadest spread of new species across the cladogram is likely to be the best to conserve because it would save a larger spread of the target organism's genetic diversity. The conservation of a broader spectrum of a group is obviously better than that of only a few closely related species. But broader conservation can be achieved only when a reliable evolutionary classification exists, so a priority for systematists is to produce the classifications upon which sound conservation can be based. The new system for the angiosperms (Chase et al. 1993; Angiosperm Phylogeny Group 1998) is a most useful tool for conservation as well as for phylogenetic classification.

Many endangered plant species have been reduced to extremely small populations, for example, island endemics in places such as Hawaii, St. Helena, or the Canary Islands. When we are faced with a very few individuals from which to recover and reintroduce a rare species, it is obviously advantageous to know details of the genetic diversity and the relationships among the individuals of the species under study. For example, the eight remaining individuals of *Alsinidendron trinerve* (Caryophyllaceae) in Hawaii were extremely closely re-

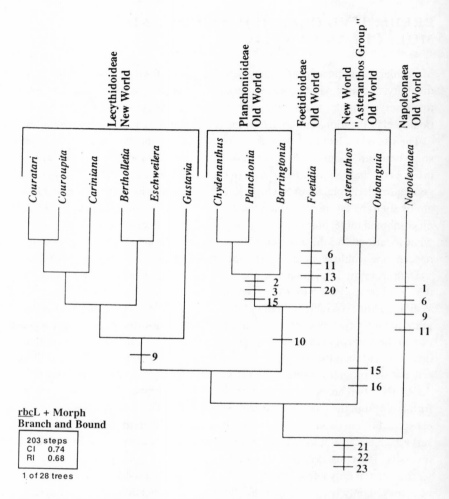

Figure 6.3. One phylogenetic tree of the family Lecythidaceae of twenty-eight equally parsimonious trees from combined molecular (*rbc*L), anatomical, and morphological data (from Morton et al. 1997).

lated. Material of *Alsinidendron* cultivated at the Royal Botanic Gardens, Kew, since it was collected in the 1950s, was from a different island and, hence, not so closely related. These plants were used for breeding to increase the genetic diversity of the population. Often the genetic relationships within a population are not as obvious as those between populations growing on different islands. In these cases molecular studies are invaluable for planning the crosses to produce the best seeds containing as much of the genetic diversity of the original population as possible. An example of this has been the work at Kew to reintroduce the lady's slipper orchid (*Cypripedium calceolus*) to the British Isles from about ten individuals known to be of wild source. The genetics of this population were analyzed to plan the breeding program. Similar work is being carried out on the toromiro tree (*Sophora toromiro*) from Easter Island and several other species, whose populations have been reduced to small numbers.

SYSTEMATIC COLLECTIONS

The greatest resources for conservation and reserve planning are the vast systematic collections of the museums and botanic gardens of the world. We need to concentrate on maintaining and using what we already have. Data from the specimens and their labels have already been used extensively to identify centers of diversity, centers of endemism, ecological conditions of an area, the onset of pollution, and much other information needed for conservation. The data contained in these collections are of much greater use when the information is available in electronic format. Although considerable effort has already been devoted to the computerization of systematic collections, progress has been slow. Large museums, such as the one at the Smithsonian Institution or the herbarium of the New York Botanical Garden, still have a long way to go. Not enough resources are being invested in databasing the collections of museums and botanic gardens on a worldwide basis or on making the data available on the Internet. Consequently, these collections are not in the best order to serve conservation.

There is an urgent need for major funding to enable the databasing of existing collections and not solely for the completion of inventory in the field. Well-databased collections that enable the manipulation of their data are among the best available tools for conservation. Since most of the large and important systematic collections are located in the developed world in northern countries, the computerization of collections has a greater urgency. It is one way in which data can be more easily repatriated to the megadiverse countries to be used both for research and for conservation. The images of type specimens now available from the national herbaria of the Netherlands, the New York Botanical Garden, and the Smithsonian are a good start, but much more needs to be done.

BANKING FOR THE FUTURE

To celebrate the new millennium the Royal Botanic Gardens, Kew, has vastly expanded its seed-banking operation to establish its Millennium Seed Bank

(Figure 6.4). The goals of this project include banking the seeds of the entire British flora by 2000 (achieved except for a few hard-to-collect rare species), banking an additional 10 percent of the world's flora by 2010, and solving some of the problems of banking recalcitrant seeds and breaking dormancy of seeds (see Prance and Smith 2000). The public display area of the Millennium Seed Bank opened in August 2000. This expanded operation has become necessary because of the number of species threatened with extinction. Two species banked in the last twenty-seven years of seed banking at Kew have already become extinct in the British Isles, and probably several others from elsewhere have met the same fate.

Seed banking is obviously a second best to in situ conservation because it does not conserve pollinators and other organisms that interact with the seed plants; however, at least it is a way of ensuring the survival of the germ plasm of many rare and endangered species.

THE ROLE OF BOTANIC GARDENS

On the whole the botanic gardens of the world have responded well to the environmental crisis and to the loss of species. Conservation and sustainable use of plants are now high on the agendas of many gardens, and Botanic Gardens Conservation International (BGCI) has played a major role in coordination of this effort over the last decade. This organization is helping to compile data on the species growing in botanic gardens, which will encourage a more coordinated approach to conservation in gardens. It has also done a great deal to provide materials for conservation education in gardens. More botanic gardens are placing an emphasis on the conservation of natural areas of vegetation within their boundaries. An excellent example is the South African botanic garden network, in which a series of gardens exist in the different vegetation zones of the country. Each garden includes areas of natural vegetation as well as the more conventional displays and collections, and a large number of South African species are preserved within the botanic gardens.

Managers of botanic gardens need to place a greater emphasis on conservation education and interpretive display of the issues involved. Their scientists must spend more time producing field guides and popular books that spread the message to a wider public audience and to more users. The field guide of the Ducke Reserve is an excellent example of a user-friendly guide to an area of tropical rain forest. The nonspecialist can easily identify plants with this guide because it is abundantly illustrated with color photographs, is laid out in easy-to-understand sections, and is based on vegetative features rather than flowers and fruit.

Figure 6.4. The palm house of the Royal Botanic Gardens, Kew, constructed 1844–48, houses many tropical rain-forest plants.

Since leaving the Royal Botanic Gardens, Kew, I am delighted to be working for the Eden Project in Cornwall, England. The whole project has been designed to promote the importance of plants to people and their sustainable use. This project, which is restoring a former china clay pit, will include a 2-hectare covered rain-forest exhibition and 0.6 hectare of Mediterranean ecosystems, as well as extensive outdoor displays of many useful plants. Like the Kew seed bank, the Eden Project seems a most suitable and appropriate way of celebrating the new millennium.

CONCLUSION

Progress has been made in legislation to control the use of, and to conserve, biological diversity. We now have the Convention on Biological Diversity, the Convention on Trade in Endangered Species (CITES), and various other pieces of international and national legislation. However, these agreements have not yet effectively slowed down species loss and fragmentation of habitat. To achieve conservation in the future, a greater emphasis must be placed on the science of conservation rather than only on politics. This will require a response from biologists. Biological scientists must devote more of their time to direct involvement in and use of their data for conservation and sustainable use of biodiversity and to education of the public. If we do not respond to this challenge, there will be no biodiversity to study and the world will face ecological disaster.

CHALLENGES FOR A NEW CENTURY

1. To resolve how many species there are on Earth through an all-species inventory
2. To convert the political debate and agenda on the various conventions governing species conservation from discussion to on-the-ground action
3. To convince the general public, through education and interpretation programs, that all life depends upon plants and that it is essential to conserve a diversity of species and to use them in a sustainable fashion
4. To complete the most accurate "tree of life" possible so that it can be employed to plan preservation of the maximum possible amount of genetic diversity

LITERATURE CITED

Angiosperm Phylogeny Group. 1998. An ordinal classification for the families of flowering plants. *Ann. Missouri Bot. Gard.* 85:531–53.
Chase, M. W., D. E. Soltis, R. G. Olmstead, D. Morgan, D. H. Les, B. D. Mishler, M. R. Duvall, R. A. Price, H. G. Hills, Y.-L. Qiu, A. Kron, J. H. Rettig, E. Conti, J. D. Palmer, J. R. Manhart, K. J. Sytsma, H. J. Michaels, W. J. Kress, K. G. Karol, W. D. Clark, M. Hedren, B. S. Gaut, R. K. Jansen, K.-J. Kim, C. F. Wimpee, J. F. Smith, G. R. Furnier, S. H. Strauss, Q.-Y. Xiang, G. M. Plunkett, P. S. Soltis, S. M. Swensen, S. E. Williams, P. A. Gadek, C. J. Quinn, L. E. Eguiarte, E. Golenberg, G. H. Learn, Jr., S. W. Graham, S. C. H. Barrett, S. Dayanandan, and V. A. Albert. 1993. Phylogenetics of seed plants: An analysis of nucleotide sequences from the plastid gene *rbc*L. *Ann. Missouri Bot. Gard.* 80:528–80.
Costanza, R., R. d'Arge, R. de Grout, S. Farber, M. Grasso, B. Hannon, K. Limburg, S. Naeem, R. V. O'Neill, J. Paruelo, R. G. Raskin, P. Sutton, and M. van den Belt. 1997. The value of the world's ecosystem services and natural capital. *Nature* 387:253–60.

Ehrlich, P. R. 1994. Energy use and biodiversity loss. *Philos. Trans. R. Soc. Lond. B* 344:99–104.

Mori, S. A., B. M. Boom, and G. T. Prance. 1981. Distribution patterns and conservation of eastern Brazilian coastal forest tree species. *Brittonia* 33:233–45.

Morton, C. M., S. A. Mori, G. T. Prance, K. G. Karol, and M. W. Chase. 1997. Phytogenetic relationships of Lecythidaceae: A cladistic analysis using *rbc*L sequence and morphological data. *Amer. J. Bot.* 84:530–40.

Myers, N. 1990. The biodiversity challenge: Expanded hotspot analysis. *Environmentalist* 10:243–56.

———. 1996. Two key challenges for biodiversity: Discontinuities and synergisms. *Biodiversity and Conservation* 5:1025–34.

Myers, N., R. A. Mittermeier, C. G. Mittermeier, G. A. B. da Fonseca, and J. Kent. 2000. Biodiversity hotspots for conservation priorities. *Nature* 403:853–58.

Nelson, B. W., C. A. C. Ferreira, M. F. da Silva, and M. L. Kawasaki. 1990. Endemism centers: Refugia and botanical collections density in Brazilian Amazonia. *Nature* 345:714–16.

Pimm, S. L. 1997. The value of everything. *Nature* 387:231–32.

Prance, G. T. 1990. Floristic composition of the forests of Central Amazonian Brazil. In *Four neotropical forests,* edited by A. Gentry. New Haven: Yale Univ. Press.

Prance, G. T., H. Beentje, J. Dransfield, and R. Johns. 2000. The tropical flora remains undercollected. *Ann. Missouri Bot. Gard.* 87:67–71.

Prance, G. T., and R. D. Smith. 2000. The millennium seed bank of the Royal Botanic Gardens, Kew. In *Nature and human society: The quest for a sustainable world,* edited by P. H. Raven. Washington, D.C.: National Research Council.

Ribeiro, J. E. L. da S., M. G. Hopkins, A. Vincentini, C. A. Sothers, A. da S. Costa, J. M. de Brito, M. A. D. de Souza, L. H. P. Martins, L. G. Logy, P. A. C. L. Assunção, E. da C. Pereira, C. F. da Silva, M. R. Mesquita, and L. C. Procopio. 1999. *Flora da Reserva Ducke: Guia de identificação das plantas vasculares du uma floresta de terra-firme na Amazonia Central.* Manaus: INPA.

Vane-Wright, R. I., C. J. Humphries, and P. H. Williams.1991. What to protect? Systematics and the agony of choice. *Biol. Conservation* 55:235–54.

Williams, P. H., D. I. Vane-Wright, and C. J. Humphries. 1992. Measuring biodiversity: Taxonomic relatedness for conservation priorities. *Australian Syst. Bot.* 4:665–79.

Williams, P. H., G. T. Prance, C. J. Humphries, and K. Edwards. 1996. Priority-areas analysis and the Manaus Workshop-90 areas for conserving diversity of neotropical plants (families Proteaceae, Dichapetalaceae, Lecythidaceae, Caryocaraceae, and Chrysobalanaceae). *Biol. J. Linn. Soc.* 58:125–57.

Wilson, E. O. 1992. *The diversity of life.* Cambridge, Mass.: Harvard Univ. Press, Belknap Press.

7

The New Age of Biological Exploration

THOMAS E. LOVEJOY

The perspectives provided by the chapters of this book are at the heart of what some have called the Century of Biology. If biological science is able to plumb the details and complexity of living systems to an extent undreamed of heretofore, then surely what lies ahead mere decades from now defies imagination. These advances will have incalculable benefit for human society that will include traditional biological sectors, such as agriculture, forestry, medical sciences, resource management, and limnology. They will also include activities such as industrial processes in which enzymes and organisms can replace toxic chemical catalysts and help in bioremediation (cleanup of wastes) and its flip side, bioconcentration (when biological processes recover and concentrate valuable but dispersed resources). The dream of industrial ecology, including nanotechnology, is much closer to realization than we think.

If biology is to be ever more central to human enterprise, the promise will rest squarely on biological diversity and our knowledge of its component parts: plants, animals, and microorganisms (Wilson 1988, 1992). If extinction rates continue to accelerate—even if they remain on the current order of one hundred times normal—the promise for science and society will be severely undercut.

It has always astonished me that many of our colleagues seem impervious to the elevated extinction rate even when, in the most self-interested sense, it represents the destruction of the library of life upon which our science is based. In contrast, when the Arno River flooded in the 1960s and the great art treasures of Florence were threatened, art historians immediately created the Committee to Rescue Italian Art. This response came not only because of the threat to the wherewithal of their discipline but also because they believed in the intrinsic importance of these resources and treasures.

The time is way overdue to get our act together as a profession and as a society. So the question is, How might we do it? How might we go beyond suffering "physics envy," bemoaning the physical scientists' ability to put together "big science" projects (with major facilities like atom smashers) and get them funded? The answer I believe lies in thinking differently about what big science is or might be. The Human Genome Project (see *Nature* Feb. 15, 2001; *Science* Feb. 16, 2001) gives us a useful point of departure, for certainly it is understood as big science by virtually everyone. Its essence can be explained in a sentence: It has a multibillion-dollar budget. At the same time it is made up of myriad increments, with each investigator driven to complete his or her share and reap the benefits as a team, and society as the biggest "winner" of all.

The Global Biodiversity Information Facility (GBIF), endorsed by the Organization for Economic Cooperation and Development science ministers in their Megascience Forum, is similarly constructed. While the goal is to have all biodiversity information (starting with data housed in and available from the great natural history museums and botanic gardens of the industrialized world) accessible to everyone via the Internet, GBIF is built up of small pieces (e.g., the Mexico specimen data, which that country has paid to have digitized in institutions like the National Museum of Natural History and the Royal Botanic Gardens at Kew). Again, it is a big science project built up of many small but significant pieces.

In a sense I am suggesting that we have fallen into the habit of not thinking big enough. When not thinking big enough, of course, one does not address questions at greater temporal and/or spatial scales or ask for enough resources to complete the tasks. And when our science questions are highly specialized, the funding response and rewards are likely to be correspondingly small.

Many years ago G. Evelyn Hutchinson (1959) asked a new question: "Why are there so many kinds of animals?" It was a question ahead of its time because we do not know how many kinds there are and cannot answer why are there so many kinds, or why are there so many of each kind. Hutchinson had important conclusions to draw nonetheless. Eugene Odum (1969) called attention to the need to investigate ecological succession (he termed this process *ecosystem development*) at a level of organization higher than the more traditional individualistic approach (Gleason 1926). This perspective resulted in new approaches, questions, and rewards. We know history has played as important a role in what is biologically possible. We also know that the interactions among life histories, biotic and abiotic processes, and biology have been both important and complex (Levin 1999). Clearly our limited knowledge of biotic diversity constrains our ability to think about such important questions as well as what they mean for a sustainable path for civilization.

We need simple themes that can encompass much of what we require and want to do. There are certainly many ways to parse these themes, but I hope the ones finally chosen will include at least elements of the following two concepts: exploring life on Earth and studying how the world works biologically.

EXPLORING LIFE ON EARTH

In the mid-fifteenth century Prince Henry the Navigator ushered in an age of discovery as he launched expedition after expedition from his observatory at Sagres to explore the world. Although they were not alone, the Portuguese held a role in exploration that was nothing sort of staggering. I believe a similar age of exploration of the biology of our planet, now possible because of abilities and emerging technologies—including examination at the level of molecules, remote sensing, and information technologies capable of taming huge amounts of information—would be an equally thrilling form of exploration, and even more likely beyond our greatest dreams.

An integrated approach would be a logical umbrella theme for systematic biology and for tackling great unknowns such as the Lilliputian world of soil biodiversity (and all its component disciplines), for example. Its scientific legitimacy rests on completing our knowledge of the basic dimensions of life on Earth. It is exciting to learn, as we have in the last twenty-five years, that entire biological communities can exist around thermal vents on the ocean floor, depending not on sunlight but rather on the primal energy of the Earth; that organisms there can exist at temperatures greatly in excess of the boiling point of water; or that seemingly inanimate particles, prions, can behave like organisms. Similar proposals have been made before (Raven and Wilson 1992). Most recently E. O. Wilson (2000) has estimated one could do the basic job for $5 billion. We need to unite as a community to seek that support, both from society and from funding agencies.

HOW THE WORLD WORKS BIOLOGICALLY

There are wonderfully exciting things to be learned about how ecosystems work and about the complex relationships between the number of species, species composition, and ecosystem function. Experimental manipulation of ecosystems, in many senses, first undertaken by the Hubbard Brook experiment (Chapter 4), can teach us many important things about natural systems and the human manipulation of them. Investigators at Hubbard Brook taught us much about the consequences of deforestation in a watershed and nutrient cycling (natural and unnatural), as well as team research. Along the way they helped us discover acid rain, or the more recent disturbing result that acid rain precipitation has leached the soil to the point where the forest seems to have stopped growing.

Long Term Ecological Research sites (LTERs), supported by the National Science Foundation, have a wealth of such insights to provide. There are important linkages and processes on even larger scales, such as how the Amazon makes half of its own rainfall, or how the floodplain forests of that greatest of all rivers provide a critical nutrient base for many fish species. These processes range all the way to the planetary scale, where we can see the metabolism of the Earth reflected in the variation in carbon dioxide concentration throughout the year.

The value of broad themes and large-scale investigations such as these is that virtually anyone can understand their needs and potential benefits, and grasp their legitimacy from a public and practical standpoint; in other words, such investigations represent the very best of integrative and applied science. Yet they also encompass a vast array of science that is curiosity-driven. Further, they provide opportunities to enhance scientific and environmental literacy, although that need could clearly stand on its own as a theme (Orr 1991).

The overall point is that by coming together in some such fashion we can be much more effective than in our current mode of operation. At the moment we are too easily dismissed, as if we were a clutch of gilded cowbird nestlings, beaks agape, chirping wildly to be fed with research dollars that we believe are some basic human right.

Organismal biologists have been figuratively "beaten into a corner," so it does not even occur to them to ask for enough. In a world where Europeans consume $11 billion of ice cream annually and Americans $8 billion of cosmetics, it is surely time to move to a more correct and larger scale for our science. Gary Barrett and Eugene Odum (1998) term this comprehensive approach *integrative science*.

In many senses ours is the kind of science — at the organismal, species, ecosystem, or landscape level — that people can most relate to. It is also obvious that our environmental footprint as a species is so great that our society must depend on joint management of the atmosphere and the biosphere. Such a challenge cannot be met without heavy investment in our (as well as other parts of) science. Consequently, as we enter the new century there is extraordinary congruence between what is right for science and what is critical for society — and it is our responsibility to do nothing less than pursue that congruence.

LITERATURE CITED

Barrett, G. W., and E. P. Odum. 1998. From the president: Integrative science. *BioScience* 48:980.

Gleason, H. A. 1926. The individualistic concept of the plant association. *Bull. Torrey Bot. Gard. Club* 53:7–26.

Hutchinson, G. E. 1959. Homage to Santa Rosalia; or, Why are there so many kinds of animals? *Amer. Nat.* 93:145–59.

Levin, S. A. 1999. *Fragile dominion: Complexity and the common?* Reading, U.K.: Perseus Books.

Odum, E. P. 1969. The strategy of ecosystem development. *Science* 164:262–70.

Orr, D. W. 1991. *Environmental literacy: Education and the transition to a postmodern world.* Albany: State Univ. of New York Press.

Raven, P. H., and E. O. Wilson. 1992. A fifty-year plan for biodiversity studies. *Science* 258:1099–1100.

Wilson, E. O. 1992. *The diversity of life.* New York: W. W. Norton.

———. 2000. A global biodiversity map. *Science* 289:2279.

Wilson, E. O., ed. 1988. *Biodiversity.* Washington, D.C.: National Academy Press.

8

Lumpy Integration of Tropical Wild Biodiversity with Its Society

Daniel H. Janzen

PHILOSOPHY

Wildland biodiversity: use it or lose it. Use it without damaging it, without lasting impact. This is biodevelopment—development of wildland biodiversity as wildland biodiversity. Be a real estate developer. Bug Acres. Swampy Hollow. Froggy Woods. Monkey Haven. Ocean View Heights. Save it, know it, and use it are the three commandments.

Conservation is place-based. Solutions are developed from the biodiversity and sociocultural traits of the place being conserved. The particular biodiversity development site I will discuss here is the Area de Conservación Guanacaste (ACG) in northwestern Costa Rica. As one moves to other tropical places, the nouns change but the actions stay largely the same—decentralization, science-based decision making, adaptive management, biodevelopment, ecosystem approach, honor thy neighbor, pay your bills on time, honest pricing. All tropical wild biodiversity is owned by some society. Integration of a conserved wildland is with that specific society, not generic society. The ACG is not a model but rather a pilot project in biodiversity-friendly tropical real estate development.

Biodevelopment of the ACG (Janzen 1999a, 1999b, 2000a, 2000b, 2002; Allen 2001) is focused on deciding what to do with a wildland once conserved, rather than on choosing a place to conserve. The conserved wildland is a sociobiophysical lump, and it always will be a sociobiophysical lump. It is a land use. And that land use must be allowed to integrate with its society if it is to persist. It must pay its bills, come to meetings on time, and send its kids to school.

This is lumpy integration. This is the gardenification of nature. It grows wild. They can be grouped as biodiversity services and as ecosystem services. The components are multicropped. They are multitasked. They have multiusers. The conserved wildland is explicitly wildland real estate development.

I do not wish to denigrate the intrinsic value and human importance of the wild biodiversity that is sprinkled throughout the urban and agricultural landscape. However, my triage view is that the great bulk of such biodiversity is fated to be tools in the toolbox, persisting at the serendipity of its direct usefulness or inconspicuousness to the owners of that landscape. Its agrobiodevelopment and its survival are another theme.

THE LUMP

The Area de Conservación Guanacaste is a UNESCO World Heritage Site of about 43,000 marine hectares and 110,000 terrestrial hectares crossing nine Holdridge Life Zones in its continuous 90-kilometer transect from marine through Pacific coastal dry forest to cloud forest (1500 to 2000 meters) to Atlantic rain forest in northwestern Costa Rica (see http://www.acguanacaste.ac.cr and http://janzen.sas. upenn.edu/caterpillars/RR/rincon_rainforest.htm). It is a decentralized portion of the Sistema Nacional de Areas de Conservación of the Ministerio del Ambiente y Energía of the government of Costa Rica, and it contains about as many species as does the continental United States (Janzen 1996).

The entrance sign to the ACG says, "Area de Conservación Guanacaste, fuente de vida y desarrollo" (Guanacaste Conservation Area, Source of Life and Development). The key word is *development,* and today this is the only national park in the tropical world with such a mission statement on its entrance. If it is to biodevelop while conserving, it must remain a lump that does not dissolve into the agroscape.

What are the biodevelopment products? We have spent 10,000-plus years making the agroscape productive, but we are still in kindergarten for the wildland garden. The ACG began its transformation only fifteen years ago (Janzen 1988). Some products, however, are already visible. And they can be formalized (e.g., Constanza et al. 1997; Daily 1997). I will outline them briefly here: biodiversity products, ecosystem products, and the megaproduct, saving the library of life.

BIODIVERSITY PRODUCTS

Ecotourism

The ecotourist is a better kind of cow. The *Guide to the Birds of Costa Rica* (Stiles and Skutch 1989) is fertilizer for the ecotourist crop. The conserved forest is the pasture. A carefully developed herd of ecotourists is easily competitive with the more traditional herd of cattle. Can they destroy a wildland? If it is small they can, just as too many cattle can trash a pasture or a water hole. To suggest that the ecotourist crop should not be developed as a wildland crop is to suggest that a dairy herd should not be maintained because too many cows can destroy a pasture. Also relevant is that humans are highly social—their impact is concentrated in only a small percent of a wildland's area, while their development benefits can be distributed throughout. And both the Smithsonian Institution and INBio's INBioParque (http://www.inbio.ac.cr) demonstrate that ecotourism need not be developed entirely on site. Lest tourism be viewed as trivial, it should be noted that it brings more foreign income to Costa Rica than do the national coffee and banana crops combined.

Bioliteracy

The bioliterate ACG neighbor pays with votes and employment capacity for the lessons learned in elementary school and high school decades earlier. The ACG annually provides a field-based biological education in the forest to all 2,500 schoolchildren living within about 20 kilometers of the area. When a child touches a harmless snake for the first time, compares leaf shapes, and counts the birds in her aural space while wearing a blindfold, she is learning to read the oldest and largest book of all. This book's content rivals that of all the world's libraries—hard copy and electronic—forever. A large conserved wildland is a Web site containing hundreds of thousands of variously integrated Web sites. I will bet any day on the bioliterate person against the bioculturally deprived, be the arena second-guessing the NASDAQ, walking a beat, or writing ad copy for iMac. The intelligent Web site will obtain the same value-added from being bioliterate. And bioliteracy just might keep the carbon-based Web site in the game a bit longer as we go about our homogenization of humanity and machine.

Bioprospecting

Biodiversity prospecting by humans (Reid et al. 1993; Dutfield 2000; Svarstad and Dhillion 2000) has been going on as long as there have been grandmothers and shamans, and other mammals and birds have long done it before us. There is no big mystery in the technology, be it blind search or biorational. The mystery lies in how to construct the income stream so that some portion of the prospecting profits are paid back to the wildland garden itself, thereby internalizing the cost of those wildlands. A one-cent charge on every cup of coffee — a rain-forest drug if ever there was one — would pay all tropical conservation costs forever. For coffee, though, it is too late to install such a payback process. The challenge in biodiversity prospecting lies not so much in more scientific knowledge but rather in building a conservation payback structure into the development of future drugs, pesticides, fertilizers, crop manipulators, perfumes, and other wildland crops.

ECOSYSTEM PRODUCTS

Water

All conserved wildlands are upstream from someone — a someone who simply must come to recognize the water factory for what it is. The ACG forested mountains provide water to more than 100,000 people. While straightforward, this factory is a hard lump to swallow. First, humanity has long viewed water as a free good, though access to it has long been a source of competitive strife. The water factory does not speak for itself; it is much in need of lawyers and accountants. Second, many of water's users dance on the very edge of their profit margins. Payment for this presumed free good will generate bankruptcies. Should these users be subsidized at the cost of eventual bankruptcy of the water factory? The tragedy of the commons raises its ugly head. Ecosystem services are particularly susceptible to such communal theft.

Carbon Farming

We all know that there is too much carbon in the air. We can stop putting it up there, and we can start pulling it down. The regenerating forest in a conserved wildland is a green scrubber for the world's smokestacks and exhaust pipes. The ACG was born with 50,000 hectares of centuries-old marginal pastures, ripe for restoration to dry forest through fire control. No, global warming will not be solved by putting forest back onto even all the tropical marginal farmlands where society is willing to grow a carbon crop (any more than national debt problems are solved with debt-for-nature swaps). However, in solving whatever small percent of the carbon problem we can with the green scrubber of forest restoration, we are capitalizing the wildland garden. Whether in a 5-ton tree or a 5-gram hummingbird, all that exquisitely formed carbon is a self-financing and self-replicating ecosystem service on which we mount a biodevelopment industry that is both further production and insurance that the carbon will stay put.

Biodegradation

The compost heap is hardly a novel concept. Give microbes their due. Biodegradation of clean agricultural waste by hungry wild biodiversity on tropical forest restoration sites is a first-class environmental service offered to the agroscape, a win-win situation. Three hectares of old pasture, nestled among ACG forest, can eat a thousand truckloads of processed orange peels in two years, and jump-start forest restoration in the process (Figure 8.1; Janzen 1999a, 2000b; Daily et al. 2000).

But when biodegradation for management and biodevelopment is conducted in a "national park" owned by a society primed for massive environmental protection, it sets up both the agroindustry and the national park to be sanctioned by a hostile urban court and competing agroindustry owners for sullying a national park (Escofet 2000). A robust government staff indeed is required to withstand the accusation that "government funds are being used to aid my competitors," and it takes a respected and science-driven conservation area staff to argue effectively that they are doing the right thing in wild biodiversity restoration and management by designing it to eat thousands of tons of agricultural waste. Janitorial custodians are not up to the task. The sense of self that is required can come about only through decentralization, something that a centralized government discourages. The self-understanding conservation area is a lump that is hard for centralized power to swallow.

THE MEGAPRODUCT: SAVE THE LIBRARY OF LIFE

The biggest product of all from a conserved wildland is simply the conservation of its wildland biodiversity, and its ecosystems, for all of humanity into perpetuity (or at least for the next generation of intelligent Web sites). This biodevelopment product is a custodial act. A conserved wildland differs from anthropogenic development in that it cannot afford to go bankrupt. We can never forget that the goal of nondamaging biodevelopment is to pay the rent and the parking tickets, to be a responsible citizen to ourselves and to the IRS, to be welcome at society's table. The ACG is the largest and longest employer in the area. However, it is not a parallel government aimed at solving all the woes of

A

B

Figure 8.1. A. One thousand truckloads of newly deposited processed orange pulp and peel deposited on April 14, 1998. **B.** The same site on December 21, 1999, after the entire mass of orange pulp and peel has been biodegraded by wild fly larvae and microbes, and the first generation of more than eighty species of broad-leafed herbs and young trees have invaded the resultant enriched soil (Modulo 2, Sector El Hacha, ACG). (Courtesy Daniel H. Janzen)

the agroscape and urbania. Its purpose is, in the limit, conservation of that which can survive in and on its lump of land use.

TOOLS FOR THE WILDLAND GARDEN

Integration of a large conserved wildland into society requires two major sets of tools—those familiar to the biologist (e.g., taxonomy, natural history, ecology, evolutionary biology, science-based decisions, agriculture, biotechnology, computerization) and those not so familiar (e.g., zoning, legislation, marketing, profit sharing, decentralization, democratization, humanity).

But even when the tools are familiar, it can be startling to use them in biodevelopment of a conserved wildland. For example, when the ACG acquires adjacent properties for rain-forest restoration, they come with old pastures intermingled with their forest patches. In contrast to the dry-forest pastures that melt in the face of dry-forest (re)invasion with the cessation of anthropogenic fires (Figure 8.2), rain-forest pastures (Figure 8.3A) persist, and persist, and persist. Commercial rain-forest pulpwood plantations, traditional demons to rain-forest conservationists, offer a tool for the elimination of these pastures (and see Parrota and Turnbull 1997). The ACG is seeding commercial gmelina plantations directly into these pastures (Figure 8.3A; Janzen 2000b). The fast-growing pulp fiber trees produce intense grass-killing shade. But shade-loving birds, bats, and small terrestrial mammals generate a steady and diverse seed rain of rain-forest shade-tolerant saplings, vines, and understory shrubs (Figure 8.3B). These understory "weeds" are then released to continue into the first stages of rain-forest restoration by harvest of the gmelina trees at the end of their (first and only) eight-year rotation. And if one is market-lucky the management endowment for the site even receives much-needed resources from gmelina sales.

The Brave New World of computerized communication has given us a tool for gentle integration of a conserved wildland with society through the Web site rather than brute integration through the chain saw and plow. We all know about Web sites. We also know about Yellow Pages. We are headed toward the melding of the functionality of the latter with the mind-blowing power of the former (Janzen and Gámez 1997; Janzen 1999a). The mechanical aspects—databases, authority files, interoperability, wireless Internet, search engines—are moving forward (and see Butler 2000). The real resource now in short supply is the very wildland information itself—images, natural history, taxonomic description—to be integrated across society's needs. If our keyboard is to do the walking, someone has to have walked the forest to fill those databases, someone has to have taken the pictures, someone has to have cleaned up the names, one or more

A

B

Figure 8.2. A. Several-centuries-old jaragua grass cattle pasture on July 25, 1972.
B. The same view on April 25, 1999, after fifteen years of elimination of manmade
fires and natural forest restoration from seeds dispersed into the site by wind and
vertebrates (Cliff Top Regeneration Plot, Sector Santa Rosa, ACG). (Courtesy Daniel
H. Janzen])

A

B

Figure 8.3. A. Rain-forest pasture planted in October 1999 with commercial gmelina trees to shade out introduced African pasture grasses (December 18, 1999; Sector San Cristobal, ACG). **B.** Rain-forest pasture after six years of unweeded gmelina plantation (the tall trees), with a dense naturally invaded understory of rain-forest tree saplings, treelets, shrubs, and vines ready to be released as rain-forest regeneration through removal of the gmelina overstory. Note the person in the lower right for scale (March 10, 1999; Rincon Rain Forest, adjacent to Sector San Cristobal, ACG, and see http://janzen.sas.upenn.edu/caterpillars/RR/rincon_rainforest.htm). (Courtesy Daniel H. Janzen)

someones will have to have been bioliterate to a degree that surpasses the writing in the *New York Times* literary section. More on this human resource to follow. Suffice to say it must be there and it must do its inventory, not so that we can count how many wildland species there are, but rather so that we can access them and understand what they do. A Yellow Pages directory is not constructed so as to be able to count the number of stores in London. The real function of bioinventory is not to select yet more sites to conserve—by and large, we already know where they are (e.g., Myers et al. 2000)—but rather to access that conserved biodiversity for nondestructive biodevelopment so that those sites are allowed to remain conserved.

The implication of the biologists' unfamiliarity with the second toolbox, the more sociological toolbox, is that (1) much novel cross-sectorial teamwork will be required for the integration of any conserved wildland with its society, (2) some conserved wildlands will fail because biologists fail to steer them away from the collision between political parties and economic factions, and (3) the most essential ingredients of all—decentralization and science-based decision making—will be slow to arrive throughout the tropics. And biologists will have to become teachers and administrators of more than graduate students and National Science Foundation grants. Bankers and bug collectors make strange bedfellows.

Costa Rica's parataxonomists and paraecologists (Janzen et al. 1993; Basett et al. 2000) are perhaps worth a case study. They are the reply to the question of who is going to gather the biodevelopment information for conserved wildlands. If there is to be a biodiversity Yellow Pages, if there is to be biodevelopment, if there is to be science-driven decision making, then some human resource must, and must quickly, make careers of being field-based bioinformation gatherers, managers, and iteration experts. Custodial management in a friendly world that highly values wildland existence for its own sake may well conserve for years and decades in the absence of understanding of what is being conserved. But the conservation area that is to survive through its biodevelopment cannot survive in ignorance. This human resource lump, this newly emerging guild at an already crowded table, must be sociologically and culturally absorbed or the conservation areas will be inoperative. However, no matter all their promise, the newly founded guild of parataxonomists and paraecologists has run awry of so many other legitimate and illegitimate social agendas that they are probably not a viable concept, except where they can be very directly nestled into larger decentralized entities that recognize their bioinventory value in spite of their conflict with established social structure. Putting administrative and scientific power in the hands of the working class is not universally welcomed in the tropics.

THE CATTLEMEN'S PARADE

The ACG has a highly visible float in the annual cattlemen's parade in the nearby provincial capital. The ACG float is made up of ACG staff—parataxonomists, program coordinators, sector caretakers, truck drivers—on horseback and carrying the national flag, the provincial flag, and the ACG flag. Cowboys in front of them, cowboys behind. Huh? Hoofprints in the forest have long been a major enemy of tropical conservation. However, the second president of the board of directors for the ACG was the president of the Provincial Cattlemen's Association (and the first president was the owner of the largest sawmill in town).

That ACG float is saying, "We are another ranch, right alongside your ranch. Our wives shop in the same stores yours do, our sons and daughters go to school with your sons and daughters, we hire your teenagers and you hire our teenagers. We are all in this together. Our products may look different from your cows, but they are, nonetheless, garden produce."

THE NEXT GENERATION

The next generation of ministers of the environment, professors of biology, and bioentrepreneurs will not come from today's parataxonomists, paraecologists, and firefighters taught that keeping fire out of the old pastures leads to tropical dry-forest restoration far faster than does planting trees. But all these adult biodevelopment managers have children and neighbors' children. These children grow up with biodevelopment managers as respected and respectable role models. Today's global conservation champions and policy makers are doing what they have been doing since childhood, just dressed in fancier clothing. From the children of those tropical paraecologists and ecotourist guides will come the next generation of tropical biodevelopment policy makers, as well as the biodevelopers in the field and forest.

SUMMARY

A conserved wildland lumpily integrated with its society is today portrayed in the Convention on Biological Diversity as "the Ecosystem Approach." The activities in the biodevelopment of the ACG from 1985 to the present, referred to earlier, may be condensed into the following summaries of this approach (from Janzen 2000c).

1. *It must be allowed to work.* Without a friendly government policy, and without people allowed to carry it out on site, conservation through biodevelopment will fail. The government policy is abetted by the global Convention on Biological Diversity and much else at the national and global levels. The people allowed to carry it out are abetted by decentralization and knowledge-based adaptive management.

2. *It is place-based.* A society has to decide what will be agroscape and urban and what will be a conserved wildland. The ecosystem approach is not so much involved with choosing where these places will be as it is focused on how a wildland will survive once designated. The willingness of society to designate is, in large part, derived from perception of land-use value to society—a value to a conserved wildland that will usually be brought about through an ecosystem approach. An ecosystem can be any size, and a given conserved wildland is likely to contain many ecosystems.

3. *It is knowledge-based.* Specific knowledge, which is largely science-based, of the place drives decisions. This knowledge—taxonomy, natural history, recovery rates, human impacts and uses, and so on— is possessed by the local human (experienced) resources (both biodiversity managers and neighbors) and possessed by society at large. Knowledge shifts and grows continually, as does the custodial challenge, leading to the essentiality of "learning by doing" and "adaptive management" toward a goal. The emphasis must be on keeping the goal at the fore and on learning the ways to that goal along the multiple possible paths. Rigid, long-term bureaucratic rules, no matter how appropriate at the moment of their invention, serve poorly as daily guidelines in this fluid biological and sociological environment.

4. *It is community-based, participatory, decentralized.* Both government and private-sector institutional and human resources can and should be full participants, but a conserved wildland also requires relinquishing of centralized political power, acceptance of local civic responsibility, honoring of biophysical boundaries, and allowing, expecting, and training of the staff of a conserved wildland to take full responsibility for it.

5. *It is designed around the organic traits of the particular conserved wildland and its local, national, and international society.* This means that each large conserved wildland will be unique in many respects.

6. *It needs to be viewed as a biophysical object unto itself rather than as an artifact of legislative action.* Actions taken need to make

biological and ecological sense, which means that relevant national laws and regulations applying to the conserved wildland may have to be far more flexible and general than is traditional in society at large.

7. *It is viewed as, and allowed to be, an entrepreneurial and directly productive sector.* It is a productive form of land use, equivalent to the agroscape in general terms. An ecosystem approach applied to a conserved wildland is not passive custody (although some conserved wildlands may exist in relatively passive custody, just as do some major art or science museums).

8. *Establishment and maintenance are optimality questions.* It must be explicitly recognized that, for example, there will always be human footprints and it is never possible to preserve "all" wildland biodiversity. Just as medicine treats a particular illness in the context of the person as a whole, a specific use of a conserved wildland needs to be viewed in the context of the entire wildland and its sociocultural placement.

9. *The conserved wildland operates under a set of rules very different from those of the agroscape.* This means that the way a species or ecosystem is treated depends on where it is encountered. A knife in the gut is a felony in one context and a lifesaving surgeon's stroke in another.

10. *Within the conserved wildland, survival of biodiversity per se, and its ecosystems, is the objective, with multiple multiused and multi-tasked by-products.* Within the agroscape and urbania, biodiversity and ecosystems are important tools in the creation and maintenance of a healthy and sane agroscape, but their survival and condition are generally not the overriding objectives, and ecosystem uses are much more monomorphic than in the conserved wildland.

11. *The conserved wildland cannot, and should not, be viewed as responsible for the environmental health of the agroscape.* However, the conserved wildland's knowledge, its human resources, and at times its actual biodiversity and ecosystem services can be very valuable ingredients for intersectorial collaborations with the agroscape and urbania.

CHALLENGES FOR A NEW CENTURY

Can tropical conservationists work with the rest of local, national, and international society to biodevelop some real and functional large conserved wildlands that are widely recognized as legitimate land use, both to inspire and to be proof of concept? Can entrepreneurs and commerce accept and develop feedback systems that ensure that some serious fraction of the budgets and profits from conserved wildland biodevelopment go to those places and their owners and managers? Can the biologists among us derive joy and a sense of accomplishment from setting up wildland biodiversity and its ecosystems for nondamaging biodevelopment by local, national, and international society? And will the governments of the countries that (still) have massive biodiversity allow all this to happen? We do not have a century to meet these challenges. We have just until tomorrow.

ACKNOWLEDGMENTS

I am not saying anything new. This is a political commentary. It is advocacy for a land-use policy. It is not a brilliant new idea. All I am doing is advocating the ideas and emotions already expressed by many people. Yet I do not cite them, credit them explicitly by name, as I would feel most comfortable doing. This is because I have found that advocating a policy is the blending of ideas, emotions, thoughts, and impressions from a blizzard of sources. As my hard disk ages I can no longer remember from whence came this or that idea. Even when I think I know, the more gentle among you remind me that it came from elsewhere. So I do very sincerely apologize to all from whom I have unabashedly taken ideas and impressions over all these years of watching the tropical forest melt in front of my eyes. I hope that your sense of my theft can be ameliorated by joy if this policy advocation does even a small amount for integration of tropical biodiversity with society.

LITERATURE CITED

Allen, W. 2001. *Green Phoenix: Restoring the tropical forests of Guanacaste, Costa Rica*. New York: Cambridge Univ. Press.
Basset, Y., V. Novotny, S. E. Miller, and R. L. Pyle. 2000. Quantifying biodiversity: Experience with parataxonomists and digital photography in New Guinea and Guyana. *BioScience* 50:899–908.
Butler, D. 2000. Search engines. *Nature* 405:112–15.
Constanza, R., R. d'Arge, R. de Groot, S. Farber, M. Grasso, B. Hannon, K. Limburg, S. Naeem, R. V. O'Neill, J. Paruelo, R. G. Raskin, P. Sutton, and M. van den Belt. 1997. The value of the world's ecosystem services and natural capital. *Nature* 387:253–60.

Daily, G. C., ed. 1997. *Nature's services: Societal dependence on natural ecosystems.* Washington, D.C.: Island Press.

Daily, G. C., T. Soederqvist, K. Arrow, P. Dasgupta, P. Ehrlich, C. Folke, A.-M. Jansson, B.-O. Jansson, S. Levin, J. Lubchenco, K.-G. Mäler, D. Starrett, D. Tilman, and B. Walker. 2000. The value of nature and the nature of value. *Science* 289:395–96.

Dutfield, G. 2000. *Intellectual property rights, trade, and biodiversity: Seeds and plant varieties.* London: Earthscan Publications.

Escofet, G. 2000. Costa Rican orange-peel project turns sour. *EcoAmericas* 2:6–8.

Janzen, D. H. 1988. Guanacaste National Park: Tropical ecological and biocultural restoration. In *Rehabilitating damaged ecosystems,* vol. 2, edited by J. J. Cairns. Boca Raton, Fla.: CRC Press.

———. 1996. Prioritization of major groups of taxa for the All Taxa Biodiversity Inventory (ATBI) of the Guanacaste Conservation Area in northwestern Costa Rica, a biodiversity development project. *ASC Newsletter,* no. 26:45, 49–56.

———. 1999a. Gardenification of tropical conserved wildlands: Multitasking, multicropping, and multiusers. *PNAS* 96:5987–94.

———. 1999b. La sobrevivencia de las areas silvestres de Costa Rica por medio de su jardinificación. *Ciencias Ambientales,* no. 16:8–18.

———. 2000a. How to grow a wildland: The gardenification of nature. In *Nature and human society,* edited by P. H. Raven and T. Williams. Washington, D.C.: National Academy Press.

———. 2000b. Costa Rica's Area de Conservación Guanacaste: A long march to survival through nondamaging biodevelopment. *Biodiversity* 1:7–20.

———. 2000c. Essential ingredients in an ecosystem approach to the conservation of tropical wildland biodiversity. Address to SBSTTA for COP 5, CBD, Montreal, Feb. 1, 2000. Unpub. ms. (www.biodiv.org/doc/sbstta/sbstta5).

———. 2002. Ecology of dry forest wildland insects in the Area de Conservación Guanacaste, northwestern Costa Rica. In *Biodiversity conservation in Costa Rica: Learning the lessons in seasonal dry forest,* edited by G. W. Frankie, A. Mata, and S. B. Vinson. Berkeley: Univ. of California Press.

Janzen, D. H., and R. Gámez.1997. Assessing information needs for sustainable use and conservation of biodiversity. In *Biodiversity information: Needs and options,* edited by D. L. Hawksworth, P. M. Kirk, and S. Dextre Clarke. Wallingford, Oxon, U.K.: CAB International.

Janzen, D. H., W. Hallwachs, J. Jimenez, and R. Gámez. 1993. The role of the parataxonomists, inventory managers, and taxonomists in Costa Rica's national biodiversity inventory. In *Biodiversity prospecting,* edited by W. V. Reid et al. Washington, D.C.: World Resources Institute.

Myers, N., R. A. Mittermeier, C. G. Mittermeier, G. A. B. da Fonseca, and J. Kent. 2000. Biodiversity hotspots for conservation priorities. *Nature* 403:853–58.

Parrota, J. A., and J. W. Turnbull, eds. 1997. Catalyzing native forest regeneration on degraded tropical lands. *Forest Ecology and Management* 99:1–290.

Reid, W. V., S. A. Laird, R. Gámez, A. Sittenfeld, D. H. Janzen, M. A. Gollin, and

C. Juma, eds. 1993. *Biodiversity prospecting*. Washington, D.C.: World Resources Institute.

Stiles, F. G., and A. F. Skutch. 1989. *A guide to the birds of Costa Rica*. Ithaca, N.Y.: Cornell Univ. Press.

Svarstad, H., and S. S. Dhillion, eds. 2000. *Responding to bioprospecting: From biodiversity in the South to medicines in the North*. Oslo, Norway: Spartacus Press.

9

Biology and the Human Sciences

Pathways of Consilience

EDWARD O. WILSON

What I have been so presumptuous, some would say reckless, as to suggest, in an age when irony and skepticism are the ruling intellectual fashion, is essentially as follows: Although it is widely assumed that there are many ways to account for the human condition, in fact there are only two. The first comes from the natural sciences, whose practitioners set out more than four centuries ago and with considerable success to understand how the material world works; and, all will agree, they have preempted that particular enterprise. The second way to account for the human condition is all the other ways.

Since the eighteenth century the great branches of learning have been classified into the natural sciences, the social sciences, and the humanities. Today we have the choice between, on the one hand, trying to make the great branches of learning consilient—that is, coherent and interconnected by cause-and-effect explanation—or, on the other hand, not trying to make them consilient. Surely universal consilience is worth a serious try. After all, the brain, mind, and culture are composed of material entities and processes; they do not exist in an astral plane floating above and outside the tangible world.

The most useful term to capture the unity of knowledge is surely *consilience*. It means the interlocking of cause-and-effect explanations across disciplines, as for example between physics and chemistry, chemistry and biology, and, more controversially, biology and the social sciences (Wilson 1998). The word *consilience* was introduced by William Whewell (1840), the founder of the modern philosophy of science. It is more serviceable than the word *coherence* or *interconnectedness,* because its rarity of usage since 1840 has preserved its original meaning, whereas *coherence* and *interconnectedness* have acquired many meanings scattered among different disciplines.

Consilience, defined then as cause-and-effect explanation across the disciplines, has plenty of credibility. It is the mother's milk of the natural sciences. Its material understanding of how the world works and its technological spin-off are the foundation of modern civilization. The time has come, I believe, to consider more seriously its relevance to the social sciences and humanities. I will grant immediately that belief in the possibility of consilience beyond the natural sciences and across to the other great branches of learning is not the same as science, at least not yet. It is a metaphysical worldview, and a minority one at that, shared by only a few scientists and philosophers. Its best support is little more than an extrapolation of the consistent past success of the natural sciences. Its strongest appeal is in the prospect of intellectual adventure and, given even modest success, the value of understanding the human condition with a higher degree of certainty.

I believe also that it is a matter of practical urgency to focus on the unity of knowledge. Let me illustrate that claim with an example. Think of two intersecting lines that form a cross, and picture the four quadrants thus created. Label one quadrant environmental policy, the next ethics, the next biology, and the final one social science. Each of these subjects has its own experts, its own language, rules of evidence, and criteria of validation. Now if we focus on more specific topics within each of the quadrants, such as forestry management, environmental ethics, ecology, and economics, we see how general theory translates into the analysis of practical problems. And we understand that in each case we somehow have to learn how to travel, clockwise or counterclockwise, from one subject to the next. In a single discussion, maybe in a sentence or two in the discussion, it is necessary to travel the entire circuit. Now in your mind move through concentric circles toward the intersection of the disciplines. As we approach the intersection, where most real-world problems exist, the circuit becomes more difficult and the process more disorienting and contentious.

The nub of the problem vexing a great deal of human thought is the general belief that a fault line exists between the natural sciences on one side and the humanities and humanistic social sciences on the other, in other words, very roughly, between the scientific and literary cultures as defined by C. P. Snow (1959) in his famous Rede Lecture. The solution to the problem, I believe, is the recognition that this boundary is not a fault line. It is not a permanent epistemological division, and it is not a Hadrian's Wall, as many would have it, needed to protect high culture from the reductionist barbarians of science. What we are beginning at last to understand is that this line does not exist as a line at all. It is instead a broad domain of poorly understood material phenomena awaiting cooperative exploration from both sides.

During the past twenty years three borderland disciplines have grown dramatically in the natural sciences, or more precisely in the biological sciences, which bridge this intermediate domain. They are, respectively, cognitive neuroscience, which is mapping the activity of the brain with increasing resolution in time and space; human genetics, including the genetics of behavior; and evolutionary biology, including sociobiology (or evolutionary psychology, as it is often called), which is tracing the biological origins of human nature. From the social sciences side the bridging disciplines include cognitive psychology and biological anthropology. To an increasing degree cognitive psychology and biological anthropology are becoming consilient with the three biology-born disciplines. In fact, they are anastomosing with them through cause-and-effect explanations. And the connections are strengthening very rapidly, as exemplified by rates of DNA sequencing and gene mapping in the human genome. Indeed DNA sequencing was essentially completed in 2001 (see *Nature* Feb. 15, 2001; *Science* Feb. 16, 2001).

Why is this conjunction among the great branches of learning important? Because it offers the prospect of characterizing human nature with greater objectivity and precision, an exactitude that is the key to human self-understanding. The intuitive grasp of human nature has been the substance of the creative arts. It has been the underpinning of the social sciences, and a beckoning mystery to the natural sciences. To grasp human nature objectively, and explore it to its depths scientifically, and grasp its ramifications, would be to approach if not to attain the grail of scholarship, and to fulfill the dreams of the Enlightenment.

Now, rather than let the matter hang in the air thus rhetorically, I want to suggest a preliminary definition of human nature and then illustrate it with examples. Human nature is not the genes, which prescribe it. It is not the cultural universals, such as the incest taboos and rites of passage, that are the products of human nature. Rather, human nature is the epigenetic rules, the inherited regularities of mental development. These rules are the genetic biases in the way our senses perceive the world, the symbolic coding by which we represent the world, the options we open to ourselves, and the responses we find easiest and most rewarding to make.

In ways that are beginning to come into focus at the physiological and even in a few cases the genetic level, the epigenetic rules alter the way we see and linguistically classify color. They cause us to evaluate the aesthetics of artistic design according to elementary abstract shapes and the degree of complexity. They lead us differentially to acquire fears and phobias concerning dangers in the environment (as from snakes and heights), to communicate with certain fa-cial expressions and forms of body language, to bond with infants, to bond con-jugally, and so on across a wide range of categories in behavior and thought. Most are evidently very ancient, dating back millions of years in mammalian ancestry. Others, such as the stages of linguistic development, are uniquely human and probably only hundreds of thousands of years old.

As an example of an epigenetic rule, consider the instinct to avoid incest. Its key element is the Westermarck effect, named after Edward Westermarck, the Finnish anthropologist who discovered it a century ago. When two people live in close domestic proximity during the first thirty months in the life of either one, both are desensitized to later close sexual attraction and bonding. The Wes-termarck effect has been well documented in anthropological studies, although the genetic prescription and neurobiological mechanics underlying it remain to be studied. What makes the human evidence the more convincing is that all of the nonhuman primates whose sexual behavior has been closely studied also display the Westermarck effect. It therefore appears probable that the trait pre-vailed in the human ancestral line millions of years before the origin of *Homo sapiens*, our present-day species.

The existence of the Westermarck effect runs directly counter to the more widely known Freudian theory of incest avoidance. Freud argued that members of the same family lust for one another, making it necessary for societies to cre-ate incest taboos in order to avoid the social damage that would follow if within-family sex were allowed. But the opposite is evidently true. That is, incest taboos arise naturally as products of response mediated by a relatively simple inherited epigenetic rule. The epigenetic rule is the Westermarck effect. The adap-tive advantage of the Westermarck effect is, of course, that it reduces inbreeding depression and the production of dead or defective children. That relentless pres-sure is almost surely how it arose through evolution by natural selection.

In another, wholly different realm, consider the basis of aesthetic judgment. Neurobiological monitoring, in particular measurements of the damping of the alpha wave, during presentations of abstract designs has shown that the brain is most aroused by patterns in which there is a 20 percent redundancy of elements or, put very roughly, the amount of complexity found in a simple maze, or two turns of a logarithmic spiral, or an asymmetric cross. It may be a coincidence that about the same property is shared by a great deal of the art in friezes, grillwork, colophons, logographs, and flag designs. It crops up again in the glyphs of ancient Egypt and Mesoamerica, as well as the pictographs of modern Asian languages.

To take the same approach but in another direction, I would like to mention biophilia, the innate affiliation people seek with other organisms and especially with the natural world. Studies have shown that, given complete freedom to choose the setting of their home or office, people gravitate toward an environment that combines three features, intuitively understood by landscape architects and real estate entrepreneurs: People want to be on a height looking down; they prefer open, savannalike terrain with scattered trees and copses; and they want to be near a body of water, such as a river or lake, even if all these elements are purely aesthetic and not functional. They will pay enormous prices for this view. People also look for two other, crosscutting elements: They want both a retreat in which to live and a prospect of fruitful terrain in which to forage, and in the prospect they like distant, scattered large animals and trees with low, nearly horizontal branches.

In short, if you will allow me to take a deep breath and then plunge where you may not wish to follow, people want to be in the environments in which our species evolved over millions of years, that is, hidden in a copse or against a rock wall, looking out over savanna and transitional woodland, at acacias and similar dominant trees of the African environment. And why not? All mobile animal species have a powerful, often highly sophisticated, inborn guide for habitat selection. Why not human beings?

And then again, in the biologically important realm of erotic aesthetics, the basis of sexual attraction, there is the matter of preferred female facial beauty open to objective analysis. The ideal subjectively preferred in tests is not the exact average, as once thought. It is not the average of faces from the general population, which can be readily blended by computer. Rather it is the average of the subset considered most attractive and then blended by computer. The ideal has higher cheekbones than the average, a smaller chin, shorter upper lip, and wider eyes, all relative to the size of the face. The evolutionary biologist might surmise that these traits are the signs of juvenescence still on the faces of young women, hence relative youth and reproductive potential. If all this seems irrational, ask any middle-aged professor whose second wife is a graduate student.

How much do we know about the innate basis of such aesthetics? Not a lot, and certainly very little about the genetics and neurobiology of the epigenetic rules—not because they have been investigated and then found lacking, not because they are too technically daunting, but simply because they have not been studied. Only recently have researchers begun to ask the right questions within the borderland disciplines.

In the creation of human nature—that is, the epigenetic rules of mental development—genetic evolution and cultural evolution have proceeded in a closely interwoven manner, and we are only beginning to obtain a glimmer of the nature of this process. We know that cultural evolution is shaped substantially by biology, and that biological evolution of the brain, especially the neocortex, has occurred in a social context. But the principles and the details are the great challenge in the emerging borderland disciplines to which I referred. In my opinion the exact process of gene-culture coevolution is the central problem of the social sciences and much of the humanities, and it is one of the great remaining problems of the natural sciences. Solving it is the obvious means by which the branches of learning can be foundationally united.

In summary, biologists, social scientists, and humanities scholars, by meeting within the borderland disciplines, have begun to discover increasing numbers of epigenetic rules such as the ones I have illustrated and speculated on here. Many more rules and their biological processes, I am confident, will come to light as scholars shift their focus to search for these phenomena explicitly.

I am very aware that the conception for a biological foundation of complex social and cultural structures runs against the grain for a lot of scholars. They object that too few such inherited regularities have yet been found to make the case solid, and in any case higher mental processes and cultural evolution are too complex, shifting, and subtle to be encompassed this way. Reduction, they say, rips human thought from its context, it is vivisectional, and it bleeds away the artist's true intended meaning. It melts the Inca gold.

But the same was said by the vitalists about the nature of life when the first enzymes and other complex organic molecules were discovered. The same was declared about the physical basis of heredity even as early evidence pointed straight to the relatively simple DNA molecule as the carrier of the genetic code. And, most recently, doubts about the accessibility of the physical basis of mind are fading before the successes of sophisticated imaging techniques. In the history of the natural sciences a common sequence has predictably unfolded as follows: An entry point to a complex system is found by analytic probing. At first one and then more such paradigmatic reductions are accomplished. Examples are multiplied as the whole system opens up and the foundational architecture is laid bare. Finally, when the mystery is at least partly solved, the cause-and-effect explanations seem in retrospect to have been obvious, even inevitable.

The value of the consilience program—or renewal of the Enlightenment agenda if you prefer a more philosophical expression—is that at long last we appear to have acquired the means either to establish the truth of the fundamental unity of knowledge or to discard the idea. I think we are going to establish it.

LITERATURE CITED

Snow, C. P. 1959. *The two cultures and the scientific revolution*. New York: Cambridge Univ. Press.

Whewell, W. 1840. *The philosophy of the inductive sciences*. London: J. W. Parker.

Wilson, E. O. 1998. *Consilience: The unity of knowledge*. New York: Alfred A. Knopf.

Index

A

acid rain, 66, 68
aesthetics, innate basis of, 152–53
algae and plant origins, 23
alien species invasions, 73–76
Alsinidendron trinerve, 121–22
annelids as vertebrate ancestors, 35–36
Area de Conservación Guanacaste, 133–43
Arno River flood, 128
Arthur, Wallace, 32–33
Atsatt, Peter, 23

B

Bachiller, Daniel, 34
bacteria: diversity of, 15–16; evolution, role in,
   9–26; nutritional modes, 19, 20–21; species
   concept and, 17–18, 19
Barrett, Gary W., xi, 1, 131
Bauplan. *See* bodies and body plans
behavior: ecology and, 89–93, 99–103, 110;
   evolution and, 89–90, 92–93, 101, 103–11;
   fear responses, 98–99; foraging, 96, 100–2;
   genetic influences on, 90–91, 94, 110; habitat selection, 102–3; multiobjective programming, 101–2; optimization model, 96;
   pattern recognition responses, 104; phylogeny and, 106–11; study, conceptual
   framework for, 91–95
Belovsky, G. E., 100
biodegradation, 137
biodevelopment, 133–46

biodiversity: challenges for a new century,
   126; conservation of, 117–26; developmental biology role in understanding, 41, 44;
   environmental services of, 120; exploration
   of, 130; Global Biodiversity Information
   Facility, 129; hot spots of, 117, 119, 120; integration with society, 133–46; inventory of,
   118–19; products of, 135–36
bioengineering, 46
biogenetic theory, 30
biogeochemistry, 58, 66, 76; readings in,
   60–63
bioinventory, 118–19, 139, 142
bioliteracy, 135
biological organization, transverse processes
   across levels of, 3
biomechanics, 38–41
biophilia, 98, 152–53
biophobia, 98–99
bioprospecting, 135–36
Blaustein, Andrew, 44–45
bmp-4 (bone morphogenetic protein) gene, 36
bodies and body plans, 28–49; limb duplication in frogs and salamanders, 41–43; vertebrate ancestors, 35–38
Borrelia burgdorferi, 69
botanic gardens, conservation role if, 124–25
Bright, Christopher, 74
brood parasitism in cowbirds, 107
Brooks, Harvey, 1